Debunking Gender Ideology and How to Manage Gender Dysphoria

by
Benjamin Bode

Dedication

This book is dedicated to my grandmother, Sonya, who helped me see writing this book to its end. It's also dedicated to all the people who feel broken beyond repair. You are not and somehow you will become whole again.

Contents

Part One: Debunking Gender Ideology

Chapter One: How Gender Ideology Attempts to Make Sex an Abstract Concept

Contents (Continued)

Chapter Two: Transition and its Negative Effects

Contents (Continued)

Chapter Three: How Gender Ideology Harms Society

Part Two: Gender Dysphoria, Its Causes and a Hypothetical Treatment Plan

Chapter One Explaining Gender Dysphoria and Its Causes

Contents (Continued)

Contents (Continued)

Foreword

While the core tenets of gender ideology have been around for far longer than the past decade, during which they have been forced into the forefront of modern western society, gender ideology is still an extremely new field of human thought, and understanding it and its associated phenomena is still developing. This text strives to be factual and provide a refutation of the tenets of gender ideology that is rooted in reality. This text is not the definitive truth, nor does it offer medical advice from a licensed healthcare professional. It is subject to being proven wrong or refuted as the understanding of the topics this text discusses deepens over time. It should also be noted that medical transition has not been studied enough to fully understand its positive and negative effects in the long term. This is especially true for puberty blockers, as much of the information that exists on them is from groups other than children with gender dysphoria who are treated with puberty blockers. Much more unbiased research must be done to fully understand medical transition and gender ideology. It must be acknowledged that this text cites several medical studies to support the presented arguments and any medical study is subject to bias from the people who performed it, along with having possibly flawed methodology that makes it unreflective of reality.

This text is meant as an overview of gender ideology and does not delve into extensive detail on all the topics it covers. The idea is to give the reader understanding of the topic as well as to present an argument to refute the core tenets of gender ideology. The text is best read knowing that each section could have its own book written about it and the understanding to do so is still developing. The reader is also advised that this book discusses topics that are sexual in nature.

Definitions

B elow is a list of definitions that are relevant to this text and understanding gender ideology:

Androphilia/Androphile: A person who is attracted to men.

Assigned Male at Birth (AMAB): A person who was born male.

Assigned Female at Birth (AFAB): A person who was born female.

Autoandrophilia: The desire to perceive oneself as male and/or to be perceived as male by society. This not only applies to physical appearance but the social and sexual functions that are stereotypically unique to men. This can be a fetishist desire.

Autogynephilia: The desire to perceive oneself as female and/or to be perceived as female by society. This not only applies to physical appearance but the social and sexual functions that are stereotypically unique to women. This can be a fetishist desire.

Asexual: One who has no romantic or sexual desires.

Binding: When a woman or girl wears a garment called a binder or uses tape to flatten her chest and hide her breasts. This is done to create the appearance of a male physique.

Bisexual: Romantic and/or sexual desire for both sexes.

Bottom Surgery: A term for surgeries that are a part of medical transition that alter the genitals to have a neutral appearance or to appear and function like the opposite sex.

Boy: Juvenile human male.

Cisgender: A term used to describe those whose gender and sex align, thus the prefix 'cis' is used which means 'on this side of'.

Chaser: Someone who pursues a relationship with a person solely due to their trans identity.

Clocking: When a person can tell that someone trying to pass as the opposite sex was not born as the opposite sex and makes it known to that person.

Deadname: When one transitions, they usually change their name. This former name is now a deadname and to use it is an act of misgendering.

Detransition: The process of stopping medical transition and returning to living life as one's biological sex.

Detransitioner: One who detransitions after being able to come to accept their biological sex during or after their transition.

Desistance: The process of ceasing social transition and returning to the identity that one had before transition.

Desister: One who desists after being able to come to accept their biological sex during or after their social transition but before starting medical transition.

Egg: A person that can become transgender but is not currently aware of it themselves or is in denial.

Eunuch: A person stripped of their sexual function though traditionally this only referred to castrated men.
Gender: The perception of one's sex.

Gender Dysphoria: A mental illness where one's perception of their sex conflicts with what their sex is in reality.

Gender Euphoria: The supposed positive feeling that comes when one with gender dysphoria is affirmed as the sex they wish to be perceived as.

Genderfluid: One who has an ever-changing gender and/or genders.

Gender Ideology: The belief that the sexual dimorphism of humans is either meaningless or non-existent and that only one's gender should be considered in all social interactions and contexts.

Genital Preference: The notion that one's aversion to performing sex acts on certain sex organs is a learned behavior that can be changed.

Girl: Juvenile human female.

Gynephilia/Gynephile: A person who is attracted to women.

Heterosexual: One who will only engage in non-platonic relationships with the opposite sex.

Homosexual: One who will only engage in non-platonic relationships with the same sex.

Hormone Replacement Therapy: The first step of medical transition and involves using hormones of the opposite sex to induce the secondary sexual characteristics of the opposite sex. Abbreviated as HRT and is also known as Gender Affirming Hormone Therapy (GAHT).

Intersex: An umbrella term used to describe a range of sex development disorders that are mostly chromosomal abnormalities that prevent one from developing fully or typically into the male or female sex.

Man: Adult human male.

Medical Transition: The medical process in which one uses various drug treatments and/or surgeries to attempt to appear and function as the opposite sex or to be androgynous.

Misgender: To identify one as the sex they were born as instead of by their gender identity.

Neopronouns: Any word that is used in place of the pronouns used to refer to people.

Non-Binary: Having a gender that is not male or female and cannot be quantified by the binary concept of sex.

Passing: The ability for one to appear as the opposite sex to most people when going about their daily life.

Rapid On-Set Gender Dysphoria: Gender dysphoria that is felt suddenly and after exposure to a social environment that encourages one to question their gender.

Sex: Trait that determines whether a person can produce male or female sex gametes. Through disease, age, medical treatments and/or deformity, one can be unable to produce gametes or carry offspring, but this does not change their sex.

Sex Assignment: The belief that sex cannot be observed as a trait at birth, because the possibility exists that the infant's gender can conflict with it and thus sex is assigned rather than observed at birth.

Sex Dysphoria: Mental anguish caused by being given secondary sex characteristics of the opposite sex when one has no gender dysphoria, or their gender dysphoria resolves after the process of medical transition.

Social Transition: The aspects of transition that don't involve medicalization.

Stealth: One who transitions to pass as the opposite sex and lives as if they aren't trans.

Top Surgery: A term for surgeries that are meant to alter the chest to appear like the opposite sex by either removing breasts or giving one artificial implants.

Trans: One who is transsexual or transgender.

Transfeminine: One who assumes a trans identity and engages in stereotypically feminine behavior.

Trans Man: A woman who identifies as a man.

Transmasculine: One who assumes a trans identity and engages in stereotypically masculine behavior.

Transmaxxing: The act of undergoing transition not due to dysphoria but to gain social benefits.

Transition: The process by which one becomes another gender.

Transgender: A term used to describe those whose sex and gender conflict and thus the prefix 'trans' is used which means 'on the other side of'.

Transsexual: One who seeks to permanently transition and live life as the opposite sex.

Transwidow/Widower: Someone who loses their relationship with their partner due to their partner's decision to undergo transition.

Trans Woman: A man who identifies as a woman

Tucking: When a boy or man uses special undergarments, tape or the anus to hide their male genitals to create the appearance of having a female pubic area.

Woman: Adult human female.

Part One:
Debunking Gender Ideology

How Gender Ideology Attempts to Make Sex an Abstract Concept

Introduction and Explaining Gender Identity

Gender ideology, the belief that people are to be seen only by the gender they are assigned at birth, has deeply harmed society and irrevocably damaged many people. It is an ideology that must be debunked, and the first step in the debunking process is to explain gender identity. The definition of gender that this book operates on is that gender is the 'perception' of one's sex, therefore a 'gender identity' is what that perception is. This means a gender identity is a mental state that is subject to change from moment to moment. It is so abstract that the term is functionally meaningless. This lack of meaning is intentional. The end goal of gender ideology is to replace the biological reality of sex and how it makes men men and women women with gender.

This attempt to replace sex with gender identity has largely failed in the eyes of most people as they think of gender as sex and can only think of two genders. However, in the eyes of legacy institutions, both owned by the state and corporations, gender identity has taken precedence over biological reality to disastrous effect, which will be explained in later sections. Society will only be willing to go along with the negative effects of gender ideology if the average person can accept that one's gender identity takes precedence over their sex. Gender ideologues desire to make sex as abstract a concept as gender identity. This is impossible, as biological reality cannot be erased. It is based on traits and characteristics that are self-evident and not feelings and perceptions that can and often do change. The first chapter will explain how gender ideology attempts to make sex an abstract concept and how those attempts fail.

Section One: Sex Assignment

Gender ideology will only have legitimacy if it can make sex an abstract concept. An important tool for gender ideologues to do so is 'sex assignment.' Sex is defined as a trait. That means it is inherent and can be observed. A person cannot assign your sex, as it is determined at the time of your conception by your chromosomes. Gender ideology makes the assertion that one who transitions to another sex has always been that sex. To explain why that person lived life as the opposite sex before transition is that their sex was merely assigned to them at birth and, as such. can change. A woman is an adult human female; that definition excludes one from being male, but gender ideology asserts one can be a woman who was assigned male at birth. This change in language makes a term used to describe one's sex abstract. If the language used to describe sex is made abstract, then the consequence is that people will begin to think about sex abstractly.

Section Two: The Gender and Sex Spectrums

A spectrum is a range of measured values that are continuous. A good way to conceptualize this is to think of sound. Humans can hear sounds ranging from a frequency between twenty and twenty-thousand hertz. This range of frequencies encompasses every sound humans can hear, from the faintest whispers shared between children on the playground to the obnoxious roar of race car engines and fireworks. The minds of gender ideologues view gender and sex through the same lens.

Gender is such an abstract concept that to claim it to be a spectrum would be logical to the average person but to claim that sex is a spectrum is ridiculous, because it functions as a binary for most of humanity. There are only men who produce sperm and women who produce eggs and have the capacity for pregnancy, childbirth and to breastfeed. While infertile men exist and there are women who lost the ability to reproduce or some and/or all of the reproductive functions associated with childbirth, this does not change the sex of either. The existence of those who are infertile along with those with disorders of sex development (DSD), also known as intersex people, are used to justify viewing sex as a spectrum. Intersex people do not change the binary nature of human sex, as everyone is male or female, because they represent such a small section of the population.

This text has defined gender as the perception of one's sex and, as such, it does not meaningfully exist, which makes its spectrum non-existent and unable to be defined. Sex is an inherent trait, and it is binary even when intersex people are taken into consideration. Gender ideologues argue that both gender and sex are spectrums that exist in tandem, and everyone falls or can fall into any number of infinite places along these spectrums. Most people are comfortable with being male or female and, while they are not perfectly masculine or feminine, they do not view themselves as being on either a gender or sex spectrum. Gender ideologues insist that they exist on both spectrums, and if they are, using the trans identity and the non-binary identity to justify the existence of these spectrums and their places within it. The next section will discuss how non-binary is used to confuse the concept of sex as well as being a leap pad to indoctrinate people into becoming gender ideologues.

Section Three: The Non-Binary Identity

A person who identifies as non-binary is someone who rejects the binary of sex and favors an identity that falls between the male and female ends of the gender and sex spectrums. This identity is entirely fabricated to justify that sex and gender both exist on a spectrum and is an identity that those who wish to be accepted by gender ideologues can adopt without expressing gender dysphoria. The identity of genderfluid is functionally the same and can be swapped with non-binary, which is what will be used going forward.

People who adopt a non-binary identity are mostly of two types: The first are people intimately invested in gender ideology and adopt a variation of a male, female or neutral identity for personal gratification. The second type are people who want to be accepted by gender ideologues but do not want to have to deal with the consequences of being transsexual or transgender. This results in most people who adopt a non-binary identity having an androgynous appearance, but are still in a state where their birth sex can be clearly identified. This makes the non-binary identity meaningless, as it literally means nothing and is just used by people who desire to structure their lives around being androgynous. Legacy institutions giving legitimacy to the non-binary identity exist as a function to confuse people and make gender and eventually sex seem like much more complex concepts than they are. The fact that non-binary people have a range of appearances and personality traits and have so many different identities are the labels that litter the gender and sex spectrum that lies concretely between male and female. The spectrum and the identities don't meaningfully exist like sex does, but they are being associated with sex and treated with the same legitimacy by gender ideologues. This act lends credence to spectrums that don't actually exist. This fails, as most people perceive themselves to be solely male or female even when they are not perfectly masculine or feminine. All the non-binary identity serves to do is to confuse and shield those who do not wish to bear the label of cisgender, which is what will be discussed next.

Section Four: Cisgender

Gender ideologues assert that sex and gender exist on a spectrum and that most people exist on the ends of that spectrum, that is, being a man who is perfectly masculine or a woman who is perfectly feminine. This is a false premise, but to legitimize it and to create an outgroup, the word 'cisgender' exists. One is cisgender if they perceive themselves to be their birth sex. This definition includes most humans, and gender ideologues use this definition to define the average human as being part of an outgroup. This serves to allow gender ideologues to shame vulnerable people into taking on a trans and/or a non-binary identity. People who accept gender ideology but do not wish to take on a trans identity can label themselves as cisgender to show solidarity and avoid some or all of the ire that gender ideologues inflict on those who are part of the outgroup. The word cisgender and the fact that there are enough people who identify with their birth sex who have taken up the label allows gender ideologues to frame gender as being real and something that every human experiences. This is of course false, as most people equate sex with gender and do not perceive their sex as deeply as gender ideologues do, nor do they associate it enough with their masculine and/or feminine traits to rationalize a transition. The word cisgender is how gender ideologues put the average human experience into the context of gender ideology; it allows them to create a large outgroup with the ingroup being the trans identified. The idea that one can be trans is false—how and why will be discussed in the next section.

Section Five: The Idea of Being Trans

There is nobody who is inherently trans. There are only people with gender dysphoria that choose to identify as either transgender or transsexual and to then undergo transition if desired. There is also a class of people who take on a trans identity to gain social benefits. These statements run contrary to the narrative that is pushed by gender ideologues who insist that being trans is a concrete and immutable identity. They argue that one's gender identity is innate and will never change once one becomes aware of it. This is false, as a gender identity is the perception of one's sex, and that can be subject to change. This change in perception can come from gender ideologues who claim to experience all, none, or some genders and can switch between them at any time. It also comes from desisters and detransitioners who abandoned their trans identity and most, if not all, tenets of gender ideology.

The existence of the detransition community disproves the justification for medical transition, as a trans identity can be transient for many and to make one undergo irreversible changes to affirm a perception that is subject to change is ridiculous. Gender ideologues deny the existence of detransitioners and they make it clear that gender is the perception of your sex and not nearly as solid of an identity as knowing your sex and acting accordingly. Gender ideology's end goal is to have one's gender identity supersede their sex. This attempt to replace sex with gender identity can only be legitimized if the idea of being trans is seen to be as legitimate as being male or female is. The process of legitimation has been described in the previous sections, but the most important is the change in language that gender ideologues have attempted to push onto society. This includes the word cisgender but also neopronouns and other terms, which will be described in the next section.

Section Six: Neopronouns and Other Language Changes

Pronouns are words that have a clear function to streamline communication and to help convey meaning more easily. English has the words 'I,' 'we,' 'he,' 'she,' 'they' 'them,' and 'it.' Pronouns were not created to affirm one's gender identity but to make English or any language more functional by being able to use them in place of proper nouns. Gender ideologues insist that pronouns are part of one's identity and they must be known and respected by everyone in society. They also insist that not using the pronouns one wishes to be used for them is an act of aggression. This means that if one refuses to refer to a man as 'she' or a woman as 'he' then that act is no different than conferring physical violence onto that person. This also applies if one refuses to refer to a person whose sex can be clearly identified as 'they' or by neopronouns, a neopronoun being any word that is used in lieu of a traditional pronoun.

The idea of treating the use of pronouns with an extreme amount of reverence serves as a function to force people to think of pronouns as synonymous with gender and then sex as gender ideology conflates the two. The intention is that people will think that you can be a they/them or an xe/xir or whatever word a gender ideologue chooses. This reinforces the gender spectrum and feeds into other language changes that gender ideologues attempt to push onto the population.

These language changes replace the words 'man' and 'woman' with terms like prostate owner and vagina haver. These terms are meant to allow gender ideologues to make claims such as: men can get pregnant or that women have penises. This is an attempt to denote the differences between men and women by making their anatomy and role in reproduction gender neutral. This fails as a biological reality and makes the differences between men and women too stark to allow the average person to rationalize them out of existence in the way gender ideologues have done. Another way gender ideologues try to skirt around biological reality is to push the concept of deadnaming, which will be discussed next.

Section Seven: Deadnaming

When one transitions, they abandon their name given at birth and take on a new name. If anybody uses their old name from that point on, the act is called 'deadnaming' and it's considered to be incredibly harmful to those who take on a trans identity. Transitioning creates a need for the people in their life to either adjust to their demands or deny them outright and, in either case, these people will deadname them either intentionally or unintentionally. This allows a trans identified person to easily isolate themselves from family and friends who refuse to support their ideology outright or are not completely on board with it, which is what those deepest in the grips of the ideology will insist is necessary to prevent harm from being inflicted onto the trans identified person. This creates an avenue to make transition more effectively maintained, as the person transitioning will only spend time with those who agree with the concept in the first place.

To forbid deadnaming also means that one must refer to and conceptualize a person's life as though they have spent it as the opposite sex for its entirety. If a man named John fathers a child and then later decides to transition into becoming a woman with the name Jessica, then we must refer to John as Jessica in every aspect of his existence. Instead of saying John is the father of one child, one must say Jessica is the mother of one child. This abandons reality, as it implies that once a transition

12

happens the identity is retroactive and that someone who is now a woman was able to have a biological child in the way a man would. The demonization of deadnaming allows gender ideologues to create friction and distance between those who disagree. It makes the concept of sex more abstract, as they can attribute sex-specific traits to someone who now identifies as the opposite sex. Deadnaming is not the only way that gender ideologues attempt to legitimize the trans identity; another way is through the concept of 'passing,' which will be discussed next.

Section Eight: The Concept of Passing

Section five of this text debunks the idea of being trans, but gender ideologues claim that a trans identity is an innate state of being. If one can transition, and society accepts that transition, then that will allow them to be their true self. Gender ideologues will state that being trans is relatively common and acceptance is all that is needed for one who is trans to live out their lives as such. This is false, as the concept of 'passing' exists, and gender ideologues will use those who take on a trans identity and pass to attempt to legitimize gender ideology. To pass is when one can look and function as the opposite sex in their daily lives and most interpersonal interactions. Gender ideologues will argue that those who pass cannot possibly be their natal sex, and many rational transsexuals wish to gatekeep the trans identity to people who invest enough into transition and are able to pass. Those who assume a trans identity and cannot pass are usually unable to engage with transgenderism to the same degree. This is in stark contrast to homosexuality; once one realizes they are gay, they can come out and live their life as a homosexual among others in their community provided they are fortunate enough to be in an accepting environment. Societal pushback against transgenderism is something gender ideologues use to push gender ideology on society. They label it transphobia and use it to make a moral argument for people to accept all the tenets of gender ideology. Transphobia will be discussed in the next section.

Section Nine: The Concept of Transphobia

Gender ideology has amassed many critics, some of which are irrational and vitriolic. These critics, along with those who calmly and rationally debunk gender ideology, are labeled as transphobic. The concept of being transphobic is used to shame people who refuse to conform to gender ideology. It can also be associated with homosexuality, as homosexuals have to tend with homophobia. This paints transgenderism as a natural phenomenon that society has not gotten around to accepting rather than a constructed ideology that harms society. The concept preys on the average person's goodwill and desire to avoid ostracization. Transphobia helped to strong-arm gender ideology into society, as there are those who do not accept the tenets of gender ideology and its end results but refuse to speak out against them out of fear of dealing with the consequences of being labeled as transphobic. Gender ideologues have attempted to legitimize the comparison of transgenderism to homosexuality by introducing the idea of 'brain sex' which will be discussed next.

Section Ten: Debunking the Idea of Brain Sex

The fundamental tenet of gender ideology is that one who takes on a trans identity is doing so to correct a mistake where their body conflicts with their gender identity. A hypothesis as to why this phenomenon occurs is that the brain sex of the individual is opposite to their natal sex (Kurth et al, 2022). Sexual dimorphism exists in the human brain to some degree (Guillamon, 2016), and some gender ideologues claim that the brains of transgender people are more in line with the sex they transition to rather than their birth sex. While medical studies using brain scanning technology have been performed, there are not only problems with these studies' methodologies but they have not reported convincing evidence to support the hypothesis that gender ideologues have asserted to be true (Luders et al, 2009) (Mueller et al, 2017).

There are issues with many of these studies; first, small sample sizes and, second, the fact that many in the transgender cohort have gone on hormone replacement therapy, which can skew results and make data inaccurate. The brain is an integral part of the body, so it is not possible for the brain to be a different sex from the body. Most scans of those with a trans identity in comparison to controls from the general population have shown that there is little, if any, difference in brain structure, especially when taking homosexuality into account (Burke et al, 2017). The brains of those are not described as being like the opposite sex but rather less like their

15

natal sex than would be typical. Many of those differences can be explained by hormone replacement therapy and homosexuality, although that is not always the case. Most of these differences exist in the right hemisphere of the brain (Zubiaurre-Elorza et al, 2012). It has also been observed that hormone replacement therapy can affect cognition. A trial was performed where men with a trans identity had better spatial awareness than women with a trans identity. Once these groups started hormone replacement therapy, their spatial awareness results fell in line with what was more typical of the opposite sex. If brains could embody the opposite sex, then the end results of this trial, after hormone therapy started, would be the same as the observations at the beginning of the study. It must be noted this study states in its introduction that conflicting evidence has been found (Nguyen et al. 2018), although it does little to detract that this evidence is a solid refutation of the argument gender ideologues use to justify transition. This supports the idea that while one's neurology can affect one's perception of one's sex, it does not inherently change it (Savic & Arver, 2011). It also most likely means that trans identifying men and women as a group have cognitive abilities that fall between their natal sex and the opposite sex, and the referenced studies had sample sizes that were too small to clearly draw that conclusion.

It also must be noted that one's social environment can influence brain development (Huang et al, 2019), and it is hard to quantify how much that affects the data being gathered, especially considering the limited number of participants most of these studies have. There are also general environmental factors to consider, and specific issues like endocrine disruptors and toxins present in the food and water supplies that most likely affect brain development. Our understanding of the brain is still extremely lacking and much more research must be performed to find out how the development of brain structure is affected by the environment and what effects brain structure has on how one perceives their sex (Wang et al, 2021).

Section Eleven: Conclusion

This text has introduced the concept of gender and how it is merely the perception of sex, and then it has established that gender ideologues have attempted to replace sex with gender. As discussed above, this attempt has failed, because they use false premises. Such premises are meant to establish the idea that gender exists and must be affirmed both socially and medically. Social transition is when one changes their name and appearance and requests that everyone must use their preferred name and pronouns. Medical transition involves the use of drugs and surgeries to imbue one with the function and secondary sexual characteristics of the opposite sex. This is done to appear and live as the opposite sex, which is known as passing, or to achieve an androgynous appearance. The next chapter will cover transition, explain social transition, and then touch on the harms of medical transition and its irreversible nature.

Chapter Two: Transition and Its Negative Effects

Section One: Social Transition and the Pipeline It Creates

Social transition is the first part of transition, and involves changes in behavior and the expected compliance of others on both a societal and individual level. This will at worst cause tension between family and peers, anguish, and trauma for the individual. These effects can mostly be reversed and most certainly be recovered from reliably. Social transition exists as a vehicle to drive an individual toward medical transition. The treatments that will be discussed in the following sections are extremely profitable, and a pipeline has been created to funnel vulnerable people of all ages into long-term medicalization. When one transitions, there is a need to pass as the sex one perceives oneself to be, and so the use of hormones to gain the secondary sexual characteristics and surgeries to mimic the sexual function of the opposite sex are insisted upon by the medical community. Medical transition is presented as the only viable treatment for gender dysphoria, and it must start with social transition. The medical community presents this treatment as an urgent case of life and death, as suicide will be the only other option for the individual if social and medical transition are denied to them. This has created a toxic environment where people who are not comfortable with the permanent effects of medical transition have to deal with such impediments. The next six sections will go into the details of medical transition.

Section Two: Puberty Blockers

If one begins medical transition during childhood then the first step will be the administration of puberty blockers. This process is marketed to children who want to transition and their parents as a 'pause button' on puberty. The medical community professes that the process can be unpaused, and puberty will play out as if they had never received the treatments with little to no irreversible side effects. This is categorically false, as puberty blockers destroy a child's developing endocrine system, and the effects of doing so are long reaching and affect the body in its totality including: the skeleton, brain, thyroid, sexual function, eyes, and digestive systems. This section will outline the irreversible damage puberty blockers cause and make a strong case to not use them as a treatment for gender dysphoria among youths.

When this text mentions puberty blockers, it is referring to a class of drugs called gonadotropin-releasing hormone agonists or GnRH agonists. GnRH agonists work by decreasing the levels of sex hormones and gonadotropins. This is done in a variety of medical cases which include: Treatment of cancers that are hormone sensitive like breast and prostate cancer, treatment of heavy menstrual bleeding and endometriosis, chemical castration of pedophiles and other extreme sex offenders, as well as to delay puberty in adolescents. Puberty is delayed in adolescents with either precocious puberty, which is when puberty begins too early in the individual and it is deemed medically necessary to be delayed, or in adolescents who have gender dysphoria and are prescribed puberty blockers to give them more time to work through their dysphoria. As stated before, puberty blockers do not simply pause puberty, they fundamentally destroy a developing endocrine system and other body systems, because artificially dropping gonadotropin and sex hormone levels can have severe side effects even when one is an adult.

The use of puberty blockers deprives one of their natural sex steroid hormones. Skeletal development is one of the many things that occurs during adolescence, and the use of puberty blockers severely stunts that necessary development, which prevents the bones from adequately collecting minerals and leads to lower Bone Mineral Density, or BMD. This most likely prevents one from reaching their natural peak BMD (Klink et al, 2015), which has severe implications for skeletal diseases like osteoporosis (Biggs, 2021). Gender ideologues that acknowledge this fact will suggest calcium supplementation to prevent one from losing BMD, but this does not compensate for the loss of function that sex hormones play in driving minerals into the bones, nor does it prevent any of the other health complications that puberty blockers cause.

The endocrine system assists with the function and development of all organs, and this includes the brain. When puberty blockers delay and/or destroy the development of the endocrine system, this stunts neurological development and can lead to complications of the brain. These complications include: pseudotumor cerebri, which is swelling of the brain for no apparent reason, and it mimics brain tumors, vision loss, nausea, double vision, headaches and cognitive decline. The FDA noticed and required labeling of these side effects after they appeared in young girls who were given puberty blockers to treat precocious puberty (Schemmel, 2022), and it's safe to extrapolate that data to children given the drugs to treat gender dysphoria. Even if mental distress and symptoms of mental illness can be abated in the short term with puberty blockers, the benefits do not outweigh the risks. It has only been shown in short term observational studies (Tordoff et al, 2022) (Chen et al, 2023), which have been criticized for drawing positive conclusions from their data (Reddit.com/r/medicine, 2023). It should also be noted that puberty blockers have been shown to cause depression in adults treated with the same GnRH agonists like puberty blockers that are used on minors (PhVWP, 2012). Even if the child has a short-term mental health gain, the long-term negative effect to neurological development should not be risked (Hayes, 2017) when these mental health issues can be treated with much less invasive methods. This is in addition to the fact that animal models show that pubertal suppression causes irreversible cognitive decline relating to memory (Hough et al, 2017). Neurological issues are not the only negative effect; as puberty blockers affect the development of the endocrine system, the thyroid is impacted as well.

The thyroid is a butterfly-shaped organ located in the neck that produces thyroid hormones that help regulate a wide range of functions in the human body. These functions range from sleep to thought patterns to sexual function and even menstruation. Puberty blockers disrupt the development and function of the thyroid. It has been observed that some children given puberty blockers have impaired thyroid function and increased levels of thyroid stimulating hormone (TSH) (Naderi et al, 2019), suggesting that puberty blockers impair thyroid development and the hypothalamus, which produces TSH, increases production of the hormone to try and compensate for the impaired thyroid function. This is evidence that indicates that puberty blockers are detrimental for the endocrine system and its development.

In addition to the general deleterious effects mentioned above, puberty blockers inhibit the sexual development of both male and female children and have a high chance of rendering them infertile. In males, puberty blockers prevent the testicles from fully maturing and producing mature sperm, while in females puberty blockers prevent menarche altogether or it prevents them from having a regular menstrual cycle that would allow for a pregnancy to occur.

Males who go on puberty blockers early enough will never experience orgasm (Cleckley, 2022). Testosterone is essential in the development of a male's sexual function, and if testosterone is not present in the body at all during a male's adolescent years, then the ability for the penis to function as a sexual organ will become impossible. Bottom surgery will not yield a functional sex organ nor imbue the male with a sex drive. Females put on puberty blockers early enough will most likely experience vaginal atrophy and pain during intercourse that makes engaging in satisfying sexual interactions difficult and in some cases impossible.

This loss of fertility is due to low levels of testosterone in males and a lack of hormone levels needed to have consistent and healthy menstrual cycles in females (Finlayson, 2016). The decreased hormone levels also affect the reproductive system in a physical manner (Smith & Urry, 1985). Females can experience vaginal dryness, bleeding, and atrophy (Yeshaya et al, year), while males experience a lack of penile development, erectile dysfunction, and low sperm count and motility.

These physical changes destroy one's ability to experience sexual function and to have a healthy sex life in adulthood, as they most likely will lack libido and their sexual organs will be unable to engage in performing sex acts effectively or at all. This becomes much more apparent for those who fully transition medically and get bottom surgery. What bottom surgery is and its consequences will be explained later in this text. Puberty blockers can and have been the first step toward one getting bottom surgery; therefore they need to be studied further to consider the full long-term effect of using these drugs on children. The physical effects of puberty blockers are not limited to the skeleton or genitals; they affect the whole body, and the eyes will be discussed next.

A cataract is a clouding of the lens inside the eye. This lens becoming clouded means that one could suffer from low vision or a loss of vision that renders them legally blind. This text mentions that puberty blockers are used to treat prostate cancer. This is one of the most common uses of puberty blockers, and as such extensive research has been done on the population of men with prostate cancer that have been treated with puberty blockers. Prostate cancer is a hormone-dependent cancer and by preventing hormone activity, the hope is that cancer growth will stop as well. One man treated with the same class of drug as puberty blockers developed cataracts (Al-Enezi 2007); this was due to prior poor vision and age. While these factors are not present in children, it is still worth noting, as puberty blockers do affect vision in relation to brain function which is something to consider as a possible long-term effect. Other effects must be considered as well like their effect on digestion, which will be discussed next.

Humans need to be able to eat and drink to live. This fact makes the ability to digest of the utmost importance. Puberty blockers have been shown to cause issues with digestion in populations of women with endometriosis and issues with infertility (Hammar et al, 2013) (Ek et al, 2015).

Puberty blockers block gonadotropin releasing hormones which have been observed to play a role in digestive function. The absence of these hormones has been shown to worsen digestion and cause issues associated with poor digestion in women with endometriosis. These symptoms include abdominal pain, bloating, constipation, nausea, and painful bowel movements. Women with endometriosis and infertile women undergoing IVF are different populations than physically healthy adolescent girls and boys; therefore it is important to keep this in mind as these findings most likely will not be generalizable to the pediatric demographic who are treated with puberty blockers for gender dysphoria. It has also been observed that puberty blockers can increase the risk of central obesity in girls that were treated for precocious puberty (Karamizadeh et al, 2013). Puberty blockers have also been shown to decrease insulin sensitivity (Nokoff et all 2021). While this evidence is concerning, more research would need to be done to see if puberty blockers administered to children increased risk of developing obesity or diabetes in adulthood.

In conclusion, puberty blockers, or GnRH agonists, are drugs that decrease the levels of sex hormones and gonadotropin releasing hormones in the body. These hormones are important for many bodily functions and artificially depleting them while one's endocrine system is developing is detrimental to their health. Puberty blockers can degrade the health of one's bones, thyroid, eyes and digestive system, as well as preventing their sexual development and rendering them infertile. The drawbacks of puberty blockers far outweigh the benefits that gender ideologues claim they confer onto children with gender dysphoria. The dysphoria and discomfort these children experience are usually natural feelings that arise as the body begins to change with puberty and one navigates the awkward transition from childhood to young adulthood. If a child is experiencing severe mental distress, then it should be seen as purely a psychological issue and treated with therapy primarily and psychiatric medications secondarily. No treatment that has the risk of permanent side effects should be administered, because it has been proven that children have a high desistance rate from gender dysphoria if they are allowed to experience their natal puberty and development (Singh et al, 2021).

Section Three: Hormone Replacement Therapy (HRT)

Subsection One: Introduction

If one transitions as an adult then Hormone Replacement Therapy or HRT is the first step of medical transition. HRT is also the next step in transition for those who took puberty blockers and look to continue with their medical transition. The goal of HRT is to change the hormone levels of one to conform to those typical of the opposite sex. This is done by giving the cross sex hormone, testosterone for women and estrogen for men, as well as other drugs that help suppress the natural hormone production in the patient. This section will cover HRT and its negative side effects in men and women in two subsections: One will cover feminizing hormone therapy for men and the other will cover masculinizing hormone therapy for women.

Subsection Two: Feminizing Hormone Therapy

Feminizing hormone therapy is the first step in MtF (male to female) transition. Estrogen is the primary female sex hormone. There are four forms, but the most potent form is called estradiol or E2. Estradiol is given as an oral pill, dermal patch, or injection along with a slew of other drugs to produce the secondary sexual characteristics typical of females in the male body. The other drugs are forms of progesterone, which is an important sex hormone that regulates a variety of functions, including ovum maturation and assisting estrogen in breast development in females, and producing testosterone in males. Antiandrogens are also used, which are a variety of drugs that prevent androgen receptors from being activated and not allowing testosterone to perform its necessary biological functions in the male body and the body in general. One of the most common antiandrogens used is spironolactone; another is cyproterone acetate. This cocktail of drugs is administered to attempt to mimic the hormone profile of an average woman ("Overview of Feminizing Hormone Therapy," n.d.). The attempt intrinsically fails, because consistent dosing of artificial hormones cannot mimic the hormone cycle that accompanies a woman's menstrual cycle. The menstrual cycle has four phases: menses, ovulation, follicular, and luteal. Hormone levels will peak, decline, and then hit a baseline so the cycle can start again. This cycle is dictated by the hypothalamus and pituitary gland in the brain. Menstruation is not a result of feminizing hormone therapy, but some who undergo feminizing hormone therapy may experience psychosomatic symptoms that conform to stereotypical symptoms of premenstrual syndrome (PMS). The decline of testosterone levels necessary to maintain proper health of the male reproductive system could also cause pain in the abdominal region. If the prostate is deteriorating due to low testosterone, then this could cause pain and amplify the psychosomatic phenomenon in some of those who undergo feminizing hormone therapy. These statements are extrapolations from the author and not definite statements of fact based on hard evidence.

What feminizing hormone therapy does achieve, in its most successful cases, is a feminine appearance: creating softer skin, gynoid fat distribution, breast development, muscle mass loss, and thinning of body hair. These cosmetic effects come at the cost of degradation of long-term health. The negative long-term health effects include blood clots, autoimmune disorders (Baillargeon et al, 2017), elevated blood pressure (hypertension), high potassium (hyperkalemia), low blood pressure (hypotension), and osteoporosis ("Treatment Risks and Side Effects of Feminizing Hormone Therapy," n.d.). Estrogen can also cause cancer, as it will stimulate estrogen-sensitive cancer cells. Osteoporosis and shortening of the spine are also possible side effects of feminizing hormone therapy. This causes one to become shorter and therefore make one perceived to be more feminine. Gender ideologues will try to construe this as a positive, but it's indicative of skeletal degeneration, which is severely harmful.

Eliminating testosterone in males stops androgenic activity and, as such, shuts down the male reproductive system. This causes erectile dysfunction and a cessation in sperm production. If a hormone environment of high estrogen and low testosterone is maintained for long enough, then complete testicular atrophy occurs and regaining sexual function becomes near impossible. The antiandrogens used are typical pharmaceuticals and have side effects reflective of that fact. High levels of estrogen and increased estrogenic activity also lower testosterone.

Progesterone is an important hormone in both men and women. For men, progesterone is used to produce testosterone, and in women it maintains libido as well as helps to regulate the menstrual cycle by allowing the ovum to properly mature and release from the ovary into the fallopian tube. The fact that progesterone helps maintain libido in women is used as a justification to prescribe it as part of feminizing hormone therapy. This is done as an attempt to mitigate and/or reverse the effects that tanking testosterone has on destroying a male's sexual function. The effect of progesterone in the context of feminizing hormone therapy is not well known. It can have limited success along with some negative side effects. It is for this reason that it is usually given only several months to several years after the initial doses of antiandrogens and estrogen are administered (Deutsch, 2016).

Feminizing hormone therapy is the first step of medical transition for many men and if successful can give one a more feminine appearance, because estrogen will force the development of female secondary sex characteristics. This comes at the cost of negative health effects from the hormones themselves as well as low levels of testosterone. Feminizing hormone therapy is far more reversible than masculinizing hormone therapy and is probably the least permanent step in medical transition. Masculinizing hormone therapy will be discussed in the next subsection; as with feminizing hormone therapy, there are some stark differences.

Subsection Three: Masculinizing Hormone Therapy

Androgens are male sex hormones. There are many but the ones most relevant to masculinizing hormone therapy are: testosterone and dihydrotestosterone (DHT). Testosterone is the primary androgen, and it converts into DHT, which is more potent, to produce the androgenic activity in males. Masculinizing hormone therapy is the first step in Female to Male (FtM) transition. It attempts to mimic the endocrine levels and environment of a typical male. This is done with exogenous testosterone, anabolic steroids, drugs to suppress menstruation, and drugs to manage DHT activity to mitigate male pattern baldness. This produces the effect of masculinization, which includes increased muscle mass, male odor, deepening of the voice, thicker facial and body hair, enlarging of the clitoris to somewhat resemble a penis, and larger extremities (Mayo Clinic, 2023) (Cleveland Clinic, n.d.). The success of masculinization varies, as testosterone can also decay into estrogen due to aromatase, which is present in fat cells. There are also genetic factors to consider. If successful, the effects of masculinization are either irreversible or difficult to reverse (Planned Parenthood, n.d.). There are negative side effects as well, which can be permanent even if reversed.

These negative side effects include vaginal and uterine atrophy, infertility, acne, pelvic pain and pain during penetrative sex and orgasm, weight gain, type 2 diabetes, blood clots in the deep veins or in the lungs and a possible increased risk of heart problems. These issues are caused by destroying a woman's natural hormone cycle by suppressing menstruation with testosterone and various birth control drugs. It is not only the increase of testosterone well above levels that a woman would be able to tolerate that leads to negative health outcomes; the lack of estrogen and progesterone is detrimental to the health of the female reproductive system and can lead to a slew of problems. In the ovaries, it can induce or exacerbate polycystic ovarian syndrome (PCOS); in the uterus, it can cause endometriosis or even uterine cancer; and, in the vagina, it can cause atrophy. Vaginal atrophy is when the muscle and tissue of the vagina deteriorate. This results in dryness and thinning of the vaginal and cervical lining. The deterioration can cause pain and be problematic on its own, but issues become apparent when a person on masculinizing hormone therapy engages in penetrative intercourse. Severe pain from overly contracted vaginal muscles and rupture of the thinned lining of the vagina can occur and result in bleeding during intercourse. Estrogen cream can be applied to the vagina and vulva

31

to prevent these negative side effects, but it can have limited success, as penetrative sex can remain painful even for women who stop masculinizing hormone therapy and detransition.

While infertility can happen because of masculinizing hormone therapy, it is possible for women undergoing masculinizing hormone therapy to become pregnant. Most women who medically transition and want to have children will actively stop taking hormones and then attempt to conceive, while a smaller group of women will become pregnant while undergoing masculinizing hormone therapy. Regardless, testosterone is toxic to a fetus and most likely causes birth defects, especially in the genitals. It is for this reason that masculinizing hormone therapy can't be administered to women who are pregnant or who are trying to conceive.

Issues with the heart are also more likely with masculinizing hormone therapy. Testosterone thickens blood, and increased muscle mass also stresses the heart more. In addition to weight gain from muscle mass, the suboptimal hormone environment that masculinizing hormone therapy creates can also lead to an increase of weight from fat. This will also stress the cardiovascular system and increase risk for heart attacks and other cardiovascular events.

While the effects of masculinization can be viewed as positive, especially if one transitions and is satisfied with the changes and feels no need to detransition, they can be negative, as the female body cannot tolerate high levels of testosterone. The most notable are growth of the clitoris and the deepening of the voice. Testosterone causes the larynx to change shape and the vocal cords to thicken. The female neck is not structured to accommodate these changes and it can cause pain, as well as removing the ability to sing. This is a change that cannot be reversed, and attempted surgical correction of the larynx would most likely cause even more damage to the voice. Testosterone causes the clitoris to grow. This is referred to as bottom growth. Bottom growth varies, so if the clitoris enlarges enough, then it can cause discomfort and chafing. Even if no discomfort occurs, bottom growth will not result in imbuing one with the functionality of a penis. While the clitoris does engorge with blood during sexual arousal, this does not function the same as an erection. There are not many cases of bottom growth that allow one to achieve penetration with a partner, and even in the cases where penetration is possible, the positions are limited. It is for this reason that many pre-op trans men will opt to use strap-ons to engage in penetrative sex. Testosterone will also result in an increased libido. This results in women who undergo masculinizing hormone therapy to experience a more male sex drive. The more violent and taboo nature of their sexual fantasies can be hard to deal with and to compartmentalize.

Masculinizing hormone therapy's effects are more permanent than feminizing hormone therapy. Androgenization is much harder to reverse than feminization, which makes detransition much more difficult for women than men, and the disruption of the complex female hormone cycle is much more complex to repair than restoring male testosterone levels to typical ranges. These effects are more pronounced for younger women and girls, as they'll be more likely to detransition as they age and develop and will have more of a challenge getting to a baseline healthy state both physically and mentally.

Subsection Four: Conclusion

Hormone Replacement Therapy is the typical first step in medical transition for men and women. The effects are varied but come with many negatives for both physical and mental health. While some effects can be reversed or mitigated, many are permanent or require a process that's too lengthy, expensive, and complicated for the average detransitioner to reverse. It is for this reason that hormone replacement therapy should only be undertaken by adults who understand the long-term health consequences and have accepted them as being worth the risk in comparison to the benefits. This remains true for all the surgical procedures that accompany a medical transition, which will be discussed next.

Section Four: Top Surgery and Binding

Men who undergo transition will usually experience some level of breast development. It is often lacking and therefore undesirable, and many opt for implants. Breast implantation is a well-established surgical procedure and process. It is also far more reversible than other procedures, as it only requires restoration of physical appearance, although there can be skin and nerve damage, which can be a challenge to recover from.

However, women who desire a male physique undergo a far more irreversible route of receiving a double mastectomy, which is the removal of both breasts. This is preceded by binding, which is when one wears a garment called a binder or uses tape to flatten the bust and create the appearance of a male physique. Binding can cause damage to the skin of the breasts, ribcage and the breasts themselves. It can cause trouble breathing and overheating when one is in a hotter climate or during the summer months. If binding is ceased and proper brassieres are worn, then these negative effects can be reversed.

What cannot be reversed is the fallout from getting a double mastectomy. A mastectomy is the next step after binding. It should be noted that binding must be stopped for a period before getting a mastectomy to allow the skin to heal.

There are five types of mastectomies: simple mastectomy, radical mastectomy, skin-sparing mastectomy, nipple-sparing mastectomy (City of Hope, n.d.) and keyhole mastectomy. A keyhole mastectomy is the least invasive form of mastectomy and involves making a small, keyhole-sized incision and removing the breast tissue through that incision. This can only be done if the skin is tight and the amount of tissue that needs to be removed is minimal (McLean Clinic, 2022). The other four types are in relation to mastectomies performed as a treatment for cancer. The amount that cancer has spread throughout the breast will determine the type of mastectomy performed. If the cancer has metastasized enough, then the entire breast is removed, along with some of the surrounding lymph nodes. This is a radical mastectomy. A simple mastectomy removes the entire breast and leaves those lymph nodes preserved. The skin-sparing mastectomy saves the skin surrounding the breast, while a nipple-sparing mastectomy preserves the nipple as well. This last type of mastectomy allows for the best results with breast reconstruction surgery in terms of preserving appearance and nerve function. This can be done with fat and skin grafts from the patient or with implants (Liu, 2021).

In the context of female to male transition, a keyhole mastectomy will be performed if the patient's breasts are small enough. If not, then either a simple mastectomy, skin-saving, or nipple-saving mastectomy will be done, depending on the patient's breast size, (Mayo Clinic, 2022) and the skin and nipples will be reattached, or the skin will be preserved, and the nipples will be tattooed on with medical tattooing. The nipples can also be removed entirely and not be replaced. Since this surgery is done for cosmetic reasons, the patient has a lot of choice over just how much of the breast is removed and how a male looking chest is constructed.

A mastectomy is a well-established procedure and if performed correctly is relatively easy to recover from. Any surgery causes the formation of scar tissue, and even a simple mastectomy will cause damage to the surrounding lymph nodes, which causes fluid buildup. As long as the fluid is drained, and the surgery site is kept clean, then recovery will happen in several weeks. If the surgery is done poorly and the chest is not drained and dressed properly, then life threatening infections can occur, along with fluid buildup. This is especially true if the patient is obese. Obesity is a risk factor for a mastectomy, and many surgeons will refuse to perform a mastectomy on obese patients.

If a mastectomy is successful, then a flat chest, which gives the appearance of a male physique, is achieved. The function of the breasts will be completely lost. The ability to breastfeed and produce breast milk will be permanently lost, as well as most sensation and the ability for the breasts to function as an erogenous zone. Some sensation may be preserved if a nipple-sparing or keyhole mastectomy is performed, but if the nipples are removed and then reattached then all sensation will be lost permanently. The scars that are formed by a simple mastectomy are usually permanent as well, and so the appearance that one once had will be extremely hard to regain, even with grafts or implants, which in themselves can have complications.

The irreversible nature of a mastectomy makes it detrimental for detransitioners to recover from and a godsend for those with gender dysphoria that persists after transition and abates because of transition. If a mastectomy is performed correctly and the patient is healthy, then the negative health effects are minimal. However, the function of the breasts will be lost, which will be an extreme burden for those who detransition and a massive benefit for those who are seeking to achieve a male physique.

While a mastectomy is an established medical procedure, what are not well-established medical procedures are metoidioplasty and phalloplasty. These surgeries are meant to attempt to imbue a female with the form and function of male genitalia and greatly miss the mark. They will be discussed in the next section.

Section Five: Metoidioplasty and Phalloplasty

Once a woman has received and recovered from a mastectomy, the next step in transition, if they desire it, is to get bottom surgery. There are two types of FtM bottom surgery: metoidioplasty and phalloplasty. However, before either bottom surgery is performed, first one would use strap-ons, packers, and stand-to-pee (STP) devices to simulate the function of male genitalia. A strap-on is a sex toy that simulates an erect male penis for intercourse. Stand to pee devices are prosthetics that allow women to urinate from a standing position. Packers are prosthetics that simulate the experience and appearance of having male genitalia. There are packers that also function as STP devices. If these devices are used and worn correctly, then adverse effects are either minimal or nonexistent.

A metoidioplasty, or meta, is a surgery where the clitoris is freed then lengthened and wrapped with skin along with the urethra. This allows for an appendage that functions somewhat like a micropenis, as it allows one to urinate from a standing position—if there are no urinary complications—and to get erections. Whether penetration can be achieved depends on the individual, but it is much more successful than if one were to attempt it with only an enlarged clitoris.

A metoidioplasty is performed purely by reshaping

the genitals to have a more male appearance and as such has far fewer complications than a phalloplasty. Complications are not nonexistent and include urethral narrowing due to scarring, urethral fistula, which is when the urethra connects with other organs, Urinary Tract Infections (UTIs), difficulties with sexual function, excessive bleeding, and difficulty urinating (Cleveland Clinic, n.d.). A scrotoplasty, which is done to shape the labia into an appendage that looks like a scrotum, can also be performed during a metoidioplasty, and testicular implants can be added as well to further mimic the appearance of male genitalia.

A phalloplasty is a much more invasive and traumatic procedure. It can be a standalone bottom surgery or performed after one has recovered from metoidioplasty. A phalloplasty consists of taking flesh from either the forearm, thigh or back, removing it, fashioning it into a tube, molding it into the shape of a penis with medical tattooing, creating a neophallus. This is then attached onto the crotch of the patient. (John Hopkins Medicine, 2023). There are many opportunities for failure with this surgery and it can even result in death. The neophallus can become dead flesh while it is being fashioned into a penis-like appendage, or die shortly after it is attached to the patient. If this occurs, then the neophallus must be removed and discarded, and a new one has to be fashioned after the patient has healed, which means another donor site would need to be used.

The attachment of the neophallus to the patient and the patient tolerating it and recovering is the first step. The second would be to lengthen the urethra so the patient can urinate standing up as if it were a natal penis. This requires the patient to use a suprapubic catheter for weeks after the surgery, and then a urologist must check if the urethra has connected to the bladder correctly to allow for urination. The third step would be to add an erectile implant and testicular implants. The last two steps of this process are optional. An erectile or penile implant is a device that resembles a tube-shaped pump that is used to replace the tissue in the shaft of the penis to cure erectile dysfunction in men that have exhausted all other options.

The complications of phalloplasty include all those that can arise during metoidioplasty, as well as wound breakdown—also known as dehiscence along the neophallus and its connection point to the body—and death of the neophallus, which can cause infection and develop into sepsis. Bladder or rectal damage can occur along with pelvic bleeding and pain. The pelvic area can also have an excess buildup of fluid that requires prolonged draining. Whether meaningful sensation beyond searing pain is preserved depends on how well the surgery was performed. The act of preserving nerves from a donor site and then attaching them to a new part on the body is complex microsurgery and extremely difficult to execute properly. The donor site can also have the complications typical of deep wounds. If a phalloplasty is done with tissue from the back, then the donor site is easier to heal. The nerves of the donor site are removed and so most, if not all, sensation is lost in the donor area as well (Osborn, 2018).

Even if a phalloplasty is executed perfectly and the

patient recovers properly over the months after the final surgical procedure is performed, the patient will not have a functional penis. They will have an appendage that can mimic an erect male penis and the urinary function of males. The patient will not be able to ejaculate or produce sperm and any erection can only happen because of an implant and not from spontaneous arousal. This makes it likely that any person who desires to interact with a penis in a sexual manner will not be satisfied with a neophallus. Orgasm with the neophallus may be possible, though, especially if the surgery is done with clitoral burying, where the clitoris is integrated into the neophallus at its base. This allows for more meaningful sexual sensation.

Both metoidioplasty and phalloplasty will usually be accompanied by removal of the female reproductive anatomy. These procedures are as follows: Hysterectomy, which is the removal of the uterus; salpingectomy, which is the removal of the fallopian tubes; oophorectomy, which is the removal of the ovaries; and vaginectomy, which is the removal of the vagina. All these surgeries are optional, and there are women who get bottom surgery while preserving their female reproductive anatomy completely, along with their fertility. If one does want to use bottom surgery to mimic the anatomy of a male, then, in the author's opinion, removal of all female reproductive anatomy is essential, especially vaginectomy. Closing off the vulva prevents standard gynecological procedures from being performed and the suboptimal hormone environment makes it more likely for cancers of the cervix and uterus to form and essentially impossible to check if those cancers are nascent. Low levels of estrogen will also cause vaginal atrophy, and if the vagina cannot be accessed to apply estrogen cream, it will lead to complications that will necessitate removal of the vagina later anyway. However, removal of female reproductive anatomy is not without risk, especially hysterectomies and oophorectomies, as it has been found that women who have undergone these procedures have an increased risk of dementia (Rocca et al, 2012). The procedures themselves are also not without complications, especially hysterectomies, which include infection, hemorrhage, vaginal vault prolapse, and injury to the ureter, bowel, or bladder. A hysterectomy can be performed either through the vagina or through an incision in the abdomen. The later method is what creates the most risk for complications. The complications arise as the procedure can damage the surrounding organs and cause them to be damaged or to prolapse into other parts of the body. Notably, the bowel can lapse into the vagina or be damaged to the point where a colostomy bag is required. These are rare complications, and the risk is decreased if the surgery is performed correctly and with the less invasive method of extracting the uterus through the vagina (Hoffmann, 2023) (Rice & Howard, 2006).

In conclusion, metoidioplasty and phalloplasty are surgeries that attempt to create male reproductive anatomy from the female genitals or donor sites on the body. Both are invasive surgeries that have complications, but phalloplasty is much more traumatic to the body and can result in death. These surgeries are usually accompanied by removal of the female reproductive anatomy which, is necessary for a more successful transition but can have severe long- and short-term negative effects on health.

Section Six: Vaginoplasty

There are many men who engage with transgenderism to satisfy a sexual fetish and thus will elect to keep their genitalia intact. However, there are men who want the sexual function of women and wish to be able to wear women's clothing without any visible bulge in the crotch. These men will undergo vaginoplasty, which is a procedure that takes the penis, testicles and scrotum and constructs an open wound that is referred to as a 'neovagina.' The neovagina is constructed by cutting open the penis and inverting it. This is called a penile inversion vaginoplasty. A peritoneal pull through vaginoplasty is performed with tissue from the abdomen at the peritoneum, and a bowel vaginoplasty is performed by using a portion of the colon to form a neovagina. Regardless of the method, the result is an open wound that requires lifelong dilation and is prone to infection and scarring.

The most common vaginoplasty performed is penile inversion vaginoplasty. This surgery involves a penectomy, which is the removal of the penis, and an orchiectomy, which is the removal of the testicles. The penis is then cut open, an opening is created between the urethra and rectum and is lined with tissue from the penis. The scrotal tissue is used to form a labia and vulva and the head of the penis is used to form a clitoris. The prostate is often moved into the neovagina to emulate

43

the G-spot of natal female reproductive anatomy. However, before this surgery can be performed, the pubic hair must be removed, and this is usually done with either laser hair removal or electrolysis. This process can cause rashes, irritation and inflammation in the skin, but the risk of complications and trauma inflicted onto the body are far lower than with a vaginoplasty.

While penile inversion vaginoplasty is the most common form of this procedure (Horbach et al, 2015), it is not the only method. A portion of the colon can be used to line the neovagina or a section of the abdomen, called the peritoneum (Mozaic, n.d.). This is done either as a revision to an initial penile inversion vaginoplasty at the request of the patient, as these methods may be more like a natal vagina than the penile inversion method or because the penis cannot provide enough tissue to successfully perform a penile inversion vaginoplasty. This occurs particularly in the case of patients who started their transition in their youth and took puberty blockers. Puberty blockers prevent a boy's penis from completely developing into adult size and as such will not give the surgeons enough tissue to work with to properly line a neovagina.

Regardless of the method used, the result is an open wound that requires lifelong meticulous care unless it is removed and sealed. The neovagina is prone to many complications, which include infection at the wound site, urinary tract infections and retention, bleeding, fluid discharge, scar tissue formation, rupture along the surgical incisions—which is known as dehiscence—and the possibility of the colon breaching the neovagina, which will cause the neovagina to leak fecal matter (Loree et al, 2020). Due to the neovagina being an open wound, it must be kept clean and dry, and it must be kept open with a process called dilation. Dilation is the process of sticking a plastic cylinder into the neovagina, which prevents the wound from closing and healing. This process starts out being done multiple times a day to only once a week. This is if recovery is optimal; a complication can arise where the neovagina heals regardless of dilation and dilation must be done daily to maintain any depth (Meltzer 2016).

To form and maintain a neovagina is traumatic to the body physically and is mentally draining, even to the point of suicide in some cases, to maintain a dilation regimen along with keeping the neovagina clean and free from infection. Another form of the procedure exists called zero or minimal depth vaginoplasty, which will just remove the penis and testicles and form a vulva with no neovagina. An orchiectomy, which is the removal of the testicles, can also be performed, and the penis left intact, which can make feminizing hormone therapy more effective. It also ensures complete destruction of male sexual function as well as making it easier to tuck. Tucking is when a man or boy hides his male genitalia by either taping it down, wearing special tucking undergarments or storing the genitals in the anus. A lack of testicles can make tucking much easier and help to relieve the dysphoria enough where a vaginoplasty is not deemed to be necessary by the patient.

A vaginoplasty is an irreversible procedure, regardless of the method used. This makes it particularly difficult for male detransitioners to recover from vaginoplasty, especially if they choose to take testosterone, as it can cause hair growth in the neovagina. This can occur even if the patient chooses to maintain transition, and along with all the other complications of vaginoplasty, which can result in death, makes this procedure one that has no medical basis for being performed. The surgery will never result in a neovagina that functions as well as natal female anatomy. The neovagina will only ever be a physical and mental burden for the patient to maintain. How much maintenance is required and how well the neovagina can function as natal female anatomy will vary depending on how well the surgery is performed and how optimal the patient's donor tissue is. To preserve sensation for the neovagina requires complex microsurgery, so to form a neovagina that has the depth for intercourse as well as the ability for the patient to orgasm requires a high degree of skill on the part of the surgeon as well as a strict regimen of dilation to maintain depth and douching to keep the neovagina from becoming infected.

A vaginoplasty is a surgery that attempts to convert male genitalia into female genitalia but can only create an open wound called a neovagina. The complications that can arise during the surgery itself or during recovery can necessitate revisions or even the removal of the neovagina. Even if the surgery is performed well and recovery is optimal, there is no guarantee that the neovagina will function well enough to offer a satisfactory amount of sexual function for the individual, or that it will function well enough in general to relieve the patient of mental dysphoria that will justify the physical and mental burden that the surgery inflicts onto them. Bottom surgeries are the most intensive transition surgeries, but there are more which will be discussed in the next section.

Section Seven: Gender Nullification and Other Transition Surgeries

Medical transition is still a new and somewhat rare phenomenon. Bottom surgeries are performed the least, but medical transition has gained traction, and there are surgeries other than the ones discussed in the previous sections that are worth mentioning. These surgeries include facial feminization surgery, feminization laryngoplasty, and another form of bottom surgery for both men and women: gender nullification surgery.

Facial feminization surgery refers to several plastic surgeries that are performed on the face to give it a more feminine appearance. This can involve breaking the bones of the face and restructuring them. While this is traumatic to the body, it is standard plastic surgery, so it is relatively easy to recover from and the operations are straightforward.

Feminization laryngoplasty refers to several surgeries that shave the larynx to remove the Adam's apple and to increase the pitch of the voice to make it sound more feminine. If this surgery is performed correctly, then it can lead to desirable results; however, it is possible that the procedure just ends with damaging the voice and not resulting in much meaningful change. What many men who wish to gain a more feminine voice do is to train their voice to have the pitch and speaking pattern of a typical woman or girl. How effective this is has many variables, but proper training can result in a man being able to sound like a woman or far more effeminate than their natal voice.

Chapter one, section three, of this text explains and debunks the non-binary identity. While the non-binary identity does not have any basis in reality, gender ideologues will insist that it must be affirmed medically with gender nullification surgery if the patient deems it necessary. This surgery is like the bottom surgeries described in sections five and six of this chapter. (Crane Center for Transgender Surgery, 2023). The key difference is that instead of trying to construct genitalia of the opposite sex, the goal is to have the crotch be as smooth and featureless as possible while maintaining urinary function. The possible complications of this surgery are like the ones described in sections five and six of this chapter. For women who wish to receive gender nullification surgery, a hysterectomy is required. Gender nullification is rare, but the surgery has been performed, unlike the surgeries the next section will cover.

Section Eight: The Theoretical Surgeries: Penis and Uterus Transplants

Sections two to seven of this text covered the various treatments that have been performed as part of medical transition. This section will cover two theoretical surgeries that have not yet been performed successfully and make a case for it to remain as such. These surgeries are uterus transplants for men and penis transplants for women.

Section five of this chapter establishes why metoidioplasty and phalloplasty mostly fail in giving the recipient of either procedure functional male genitalia. There is acknowledgement of the pitfalls of these surgeries, and some have proposed the idea of penis transplants as an alternative. This surgery has been performed once on a man who was severely wounded in Afghanistan and lost both his legs. The surgery involved taking a donor penis and the surrounding abdominal tissue and attaching the arteries, nerves and blood vessels to the patient. The testicles were not attached to the patient, as it would result in the patient being able to reproduce with the donor's genetics instead of his own. This procedure was fourteen hours long and involved complex microsurgery on a patient who had a pubic area that was analogous. (Nitkin, 2018). Doing this on a woman would certainly present its own challenges and complications. It would also be unethical to perform a testicle transplant for the same reason it was not performed on the injured man. Therefore, even if this surgery is performed successfully on a woman, it would not give her male reproductive function. Add the fact that most transplants require the patient to take immunosuppressant drugs makes this surgery even less feasible.

Transplants have been performed on women. These women have gone on to have successful pregnancies with IVF and then will have the uterus removed so they no longer need to suppress their immune system to prevent their body from rejecting the transplanted uterus. The uterus does not exist in a void, and neither does any organ. The uterus will not be able to function properly if it is not connected to the various ligaments and connective tissue that keep it stable and healthy in a woman's body. While it is possible to perform this surgery on a man, it has never been shown that this surgery can result in a successful pregnancy. The first known male recipient of a uterus transplant was Lili Elbe, and while he did survive the implantation, his body rejected the uterus and he died of cardiac arrest during subsequent complications. Modern medicine would allow a man receiving a uterus to tolerate the surgery better, but a pregnancy would be a disaster. The expansion of the uterus as the fetus develops would destroy the surrounding organs and muscles in the male body and cause death to both the fetus and man. Add that artificial hormones would most likely have detrimental effects on the development of the fetus and the male's endocrine system and pituitary gland would have no means of maintaining the hormone levels and producing the hormones needed to have a successful pregnancy. Uterus transplantation is something that exists strictly as theoretical in terms of performing the surgery in a man and having a pregnancy in the way that women who have received the procedure do so.

Medical transition in its current form will never allow a person to gain the reproductive function of the opposite sex, and that may remain the case regardless of the advancements in medical technology that are sure to come about. While sections two to seven of this text make a strong case for one to never undergo medical transition, the next section will lay out guidelines for administering medical transition as well as to conclude this chapter of the text.

Section Nine: Conclusion and Guidelines for Medical Transition

There is no medical basis for medical transition. These procedures are given only to relieve mental discomfort that is called gender dysphoria. Gender dysphoria can be so severe that it can push one who experiences it severely enough to suicide. The medical community presents medical transition as the only way to alleviate this suicide ideation. There is little hard evidence of this being the case. It should be noted that the suicide rate in children is low (Biggs, 2022) and that it has been observed that those who go through medical transition are at higher risk for suicide (Dallas et al, 2021) (Dhejne et al, 2011). The surgeries and hormone treatments can inflict many complications and great pain, along with the fact that underlying mental health issues may go untreated in preference for addressing gender dysphoria. The greatest risk of transition would be the development of sex dysphoria. This is the phenomenon when one develops secondary sexual characteristics of the opposite sex and feels great discomfort as a result. A notable example is Alan Turing, who was criminalized for homosexuality and forced to take estrogen as a form of chemical castration to avoid prison. The ensuing breast development and impotency may have been a contributing factor in his suicide. Sex dysphoria can only develop either by transitioning and then choosing to detransition due to accepting your natal sex or being forced into some or all the medical procedures that constitute medical transition.

This chapter of the text presents a strong case against medical transition. However, there will always be those who seek access to medical transition. It is also the opinion of the author that a full-on ban of medical transition would be ineffective, and it would be far greater for society if the medical community adapted the following guidelines:

1. Under no circumstances is anybody under the age of twenty-one allowed to receive any treatment that is part of medical transition. This would cease the practice of using puberty blockers. Any child that feels distress due to gender dysphoria should receive psychiatric care with a focus on alleviating gender dysphoria by accepting their natal sex.

2. The process of medical transition should only be started after a year of psychiatric care is given with a focus on treating any possible mental health comorbidities along with attempting to relieve gender dysphoria, and the patient should have informed consent before any treatment is given.

3. Focus on reducing the physical trauma of medical transition as much as possible. If a bottom surgery must happen, then for men an orchiectomy should be performed before any vaginoplasty proceeds, and for women only metoidioplasty should be performed and phalloplasty should be done away with entirely or at least highly discouraged.

If the medical community adopted these guidelines, then it would protect children as well as vulnerable gender non-conforming adults. It would also make transition far less harmful for the individual. However, the current practice of medical transition is deeply harmful to the individual, and it has only reached its current state because gender ideology has been pushed onto society enough to cause deep harm. The next chapter will explain gender ideology's negative impact on a societal level.

Chapter Three: How Gender Ideology Harms Society

Section One: Introduction and Overview of How Gender Ideology Harms Society

Chapter two of this text delved deep into medical transition, which is how gender ideology causes harm to the individual, while this chapter will explain how gender ideology harms society. The first section will serve as an introduction and give an overview of such harm. The next three sections will explain how gender ideology harms women, children and men.

First, there are medical resources that are dedicated to medical transition that could be used for actual constructive healthcare. This includes: The hospital space and staff, along with surgeons and possibly even blood supply, and the various drugs used as part of medical transition often have other applications that are necessary for maintaining the physical health of patients and the demand that medical transition creates for these drugs can cause shortages. Second is that academics and policy setters in the medical community spend resources on studying and setting guidelines for medical transition instead of using their time and focus on other fields of study. Lastly, there are gender clinics that are established and staffed that could be healthcare clinics for other needs. The staff in gender clinics and in other healthcare settings choose to specialize in pushing medical transition instead of specializing in other areas of healthcare that would be constructive to society.

It should be noted that pushing gender ideology onto society exacerbates the polarization that it has experienced. A minority of vehement ideologues who have infiltrated academia and other legacy institutions to push this issue, along with those who push back against them, creates strife and conflict in society when an environment for collaboration and mutual understanding could be created instead.

This is a brief overview of some of the societal issues that gender ideology has created. The next three sections will dive into how gender ideology has created issues for men, women and children on a societal level.

Section Two: How Gender Ideology Harms Women

Women have historically had to deal with sex-based discrimination, and while much progress has been made to alleviate the societal ills they have faced, modern society has created problems for women and gender ideology is the source of the following problems for women: The inability for lesbians to set sexual boundaries; the infiltration of men into women's sports, prisons and bathrooms; dehumanization of women; the use of transition to treat sexual trauma instead of psychiatric care; and other negative effects on women's healthcare, including the destruction of some women's heterosexual relationships.

The homosexual community has fought for decades to be able to form same-sex relationships without being criminalized and stigmatized by society. Gender ideology has done a great disservice to the homosexual community, notably the lesbian community. Many straight men engage with transgenderism to satisfy a sexual fetish. These men will call themselves lesbians or transbians and pursue relationships with actual lesbians. This not only creates an environment for sexual assault, but many gender ideologues will shame lesbians when they refuse sexual and romantic advances from these men. What gender ideologues use to shame those attempting to set sexual boundaries is the concept of genital preferences. A genital preference is that one can choose what sex they are attracted to. Therefore, if a lesbian refuses to have sex with a man that claims to be a woman, it can be seen to be the result of learned behavior that can be changed and not her exercising her sexuality. If lesbians refuse to acknowledge they have a genital preference and endeavor to change it, then they are transphobic in the eyes of gender ideologues and must be met with shame and ire. This creates a culture of fear and intimidation in the realm of dating. This means that women are more likely to be victims of sexual assault by men and gender ideologues will have an interest in covering up these instances of sexual assault of lesbians from trans-identified men to prevent even more societal pushback on gender ideology.

Sex segregation primarily benefits women. Generally, women are physically smaller and weaker and therefore are more likely to be victimized if natal men are allowed into women's bathrooms, prisons, and locker rooms. The close quarters and more intimate nature of these environments makes it much easier for men to assault women, especially when they have access to these environments in the same way women do. There is also the area of women's sports to contend with. The typical man is more inclined to athleticism than the typical woman, and this trend does extrapolate to the highest tiers of athletic skill in most cases. This means that men who are mediocre athletes in men's divisions have an extreme advantage when competing against natal females for many reasons. This, added with the social benefits that gender ideologues can give to these men, has comprised women's sports in many areas and pushed women out of achieving athletic titles and having careers as athletes due to the inclusion of men in their leagues, divisions, and competitions.

Women have always had to deal with those who would reduce them to their biological function and gender ideologues do this in two insidious ways. The first is to insist that one does not need to be born female to be a woman and so women can have penises or vaginas in the mind of gender ideologues. They will insist on calling women birthing people, bleeders, menstruators and vagina owners, instead of mothers, women, or girls, when referring to biological functions unique to women as to not detract from the supposed innate and legitimate womanhood that trans identified men experience. This serves only to degrade women's standing in society as well as to legitimize the delusion of trans-identified men, whether that be as the result of a fetish or the desire to be a straight woman instead of a gay man or a complication of mental health and responses to trauma.

Women's healthcare has historically been

underfunded and understudied, which leaves a lot of women to suffer at the hands of a medical system that is ill-equipped to deal with their problems. Gender ideology consumes resources of the medical community, which prevents this problem from being corrected. It also prevents the medical community from acknowledging the biological reality of women, which makes it harder for the medical community to gain the understanding of female biology necessary to be able to administer proper healthcare for women. Medical transition being pushed onto women as a solution to heal from sexual trauma is another actor to consider. The second part of this text will dive deeply into this, but there are women and girls who have been sexually assaulted and raped and experience gender dysphoria to disassociate from their trauma and as a method to prevent future assaults from happening to them. If these women transition instead of receiving empathetic and quality psychiatric care for their trauma, then it will never truly be resolved and only leave the women to be victims to the many health complications the second chapter of this text describes.

There are heterosexual men that engage with transgenderism to fulfill a fetish and/or because of a porn addiction and being exposed to the social contagion aspects of gender ideology. These men often have wives and children. Their wives become transwidows as the man that was their husband is now a trans woman. This can create an abusive environment for the women involved in these relationships. These women will often lose attraction to their partners as they adjust to the new identity, and the proposed new sexual dynamics are too much for their relationship to weather. These relationships often do fall apart after the male transitions, which creates strife and degrades the quality of life for the women and children that are a part of their families. There are also lesbian relationships that are broken due to one partner pursuing transition. This is in addition to masculine women that transition and seek to pursue relationships with women as if they were straight men. These issues in dating affect women far less than straight men that pursue a trans identity.

Women have a unique role in society and as such have had their own unique problems to deal with in navigating their world. Gender ideology has added to their problems and has made it easier for women to be assaulted in women's bathrooms, prisons, and locker rooms. They have also made women's sports open to being dominated by opportunistic men, dehumanized women, and presented even more issues in women's healthcare. There are women who have lost their partners and stable families due to their partner's decision to transition. These are some of the ways that gender ideology negatively affects women. The next section will explain the negative effects men face because of gender ideology.

Section Three: How Gender Ideology Harms Men

There are those in society that consider men to be free from societal problems, but this is not the case. Men have to deal with societal pushback, and gender ideology has added onto that pushback. These problems include men being presented with transition instead of self-improvement to elevate their standing in society and creating a violation of sexual boundaries.

Being a man that is considered traditionally masculine is not an easy endeavor. This is especially true in a society that discourages such behavior, like in modern Western nations. There are many men who feel that they are not good enough to be well-off in society and that they can never improve themselves enough to hit that mark. Instead of improving themselves while accepting their natal sex, a portion of these men have resolved to transition and to use gender ideology to improve their standing in society. This phenomenon is known as transmaxxing. Transmaxxing allows these men to experience intimacy with women or even other men and to have a much more active social life. While this usually is just love bombing and gaslighting, the social interaction is more than enough to keep these formerly isolated men captured in their trans identity instead of abandoning it and forging a strong masculine identity, an identity that would lead to long-term happiness and fulfillment.

Gender ideology has also violated the sexual boundaries of men, which happens mostly to straight men. There are gay men that cannot accept that they are homosexual and endeavor to live their lives as straight women. Some of these men will be deceitful or even use aggressive tactics to engage in intimate relations with straight men. Others in this community will endeavor to form more healthy relationships but often fail. Gay men can also experience a violation of their sexual boundaries as women who pursue a trans identity can and will seek relationships with gay men that have dynamics they're inherently uncomfortable with. This discomfort can be invalidated and shamed with the concept of genital preferences. It is analogous to what gender ideologues do to lesbians who refuse to engage in intimate relations with men that pursue a trans identity.

The problems that men face because of gender ideology may be different from women; they can be just as devastating given the right circumstances. Violating sexual boundaries is often traumatic, and anything that can create an environment where that is more likely to happen should not be encouraged. This problem is in addition to the fact that many men abandon accepting reality and working to improve it, which is necessary to achieve true self-actualization, in preference for pursuing a trans identity. While these issues are not the only things that affect men, they do nothing to improve the lives of men or to create a better society. They do quite the opposite. Gender ideology also has a similar and disastrous effect on children, which will be explained in the next section.

Section Four: How Gender Ideology Harms Children

Children are the most vulnerable members of society and as such require the greatest amount of protection. Gender ideologues are aware of the vulnerability of children and use that to push their agenda as well as to abuse them for their own gratification (Gluck, 2022). The issues that gender ideologues force onto children include removing children from parental or guardian custody in situations where they are safest, creating a pipeline to justify the sexual abuse of children, preventing children from maturing, and allowing them to engage in gender non-conforming behavior.

A child is often the safest with their parents or whoever has custody over them. Gender ideologues create an environment that makes it more likely for parents and guardians to be labeled as abusive parents. This is done by claiming the parents do not accept their child's trans identity and/or actively restrict their ability to transition. This justifies having children be taken into state custody. This makes it more likely for a child to be abused, as many orphanages and foster homes are run poorly. This issue compounds when children who are separated from their parents or guardians based on them not being allowed to pursue a trans identity are placed into sex-segregated spaces that are opposite to their natal sex. A girl in the boy's section of an orphanage or group foster home is much more likely to be abused and assaulted. This adds more trauma that could have been avoided if gender ideologues did not force the state to intervene in helping certain children pursue trans identities.

The trauma and abuse that gender ideologues have inflicted onto children is most likely of little concern to them, as another goal of theirs is to justify the sexual abuse of children. The process of inflicting medical transition onto someone, especially a child, is done to satisfy a fetish for some who push gender ideology. This abuse also coincides with sexual abuse of children, and these activists of gender ideology wish to use transgenderism and medical transition as a vehicle to make pedophilia seen as at least tolerable in society. The surgeries described can be referred to as sex reassignment or gender reassignment surgery. If a child can consent to changing their sex or gender, which gender ideologues conflate with sex, then it will not take much to make the public opinion on what is acceptableaccommodate the viewpoint that children can consent to sex. Many children have also been introduced to sexually explicit material like the graphic novel, Gender Queer, in public schools as a method to push gender ideology onto them. Sexual trauma experienced during childhood often stunts development and the ability for children to mature into mentally healthy adults. Section two of the second chapter of this text describes the effects of puberty blockers and while there are many effects, one of the most devastating is that they never allow a child to develop properly as an adult. Puberty blockers prevent a child from reaching true physical and mental maturation, which guarantees that they will never be fully functional adults. Fewer functional adults means that we will have a less functional society.

There are children that go against the grain and engage in behavior that is untypical for their sex. These feminine boys and masculine girls are prime targets for gender ideologues. They will single out these children and push the idea onto them that they cannot engage with these behaviors and still accept their natal sex. Their interests mean they have a trans identity that must be affirmed socially and then medically. Many of these children would grow up to be homosexual or abandon their sex untypical behavior when they hit puberty. Regardless of the outcome, gender ideologues rob these children of experiencing their natural development for daring to not conform to societal norms.

To conclude, children are the most vulnerable members of society. Gender ideologues are aware of this and prey on children, as they are the easiest to convert into devout defenders and members of gender ideology. Teaching children that they are transgender and allowing them to take on a trans identity is bad for their mental health (Heino et al, 2023). It also puts these children on a path to social upheaval of their lives and medicalization, much of it ending with mutilation, as detailed in chapter two of this text. The push to normalize gender ideology among the youth has also led to an environment where children are exposed to explicit sexuality as well as being made to endure one of the most egregious forms of sexual abuse, which is going through medical transition as a child. Gender ideology has funneled kids into medicalization and forced kids who do not conform to strict gender roles to be a part of a deeply damaging ideology. The harm gender ideologues have inflicted onto children will be felt for generations to come.

Section Five: Conclusion to How Gender Ideology Harms Society

The push to normalize the tenets of gender ideology among the general population has had disastrous effects on men, women, and children. The violation of sexual boundaries of both men and women has been disastrous not only for heterosexual relations but especially for the homosexual community. Not only have gender ideologues groomed children that would grow up to be homosexual adults into taking on a trans identity, their insistence on pushing the idea of genital preferences onto society and associating transphobia with homophobia has been a major setback for the homosexual community and the acceptance of same-sex relationships.

Gender ideologues will either outright deny that gender ideology has caused any harm to the individual or society or will insist that any collateral damage and harm is necessary to alleviate and treat gender dysphoria in individuals that present with it. The next part of this text will dive into gender dysphoria. It will explain what it is and why gender ideologues miss the mark on exactly what it is and how best to treat it.

Part Two:
Gender Dysphoria,
Its Causes, and
a Hypothetical
Treatment Plan

Chapter One: Explaining Gender Dysphoria and Its Causes

Section One: Introduction and Explaining Gender Dysphoria

Gender dysphoria is not an innate state of being that must be affirmed and treated with medical transition. It is a mental illness where one's perception of their sex conflicts with their sex in reality. This mental illness, like any other, is complex and can be caused by many factors that can and often do interact with each other. Gender dysphoria ranges greatly in severity. It can be a mild inconvenience or so debilitating that it burdens one who suffers with it with daily suicidal ideation. While gender ideologues will acknowledge the mental distress that results from gender dysphoria, they will refuse to call it a mental illness. Instead, they insist that gender dysphoria results because one has a gender identity that conflicts with their assigned sex at birth and that is not affirmed socially and with medical transition. This is false, as gender dysphoria is a mental illness that has causes typical of other psychiatric disorders and can be resolved or at least managed with psychiatric care. When gender dysphoria was considered a mental illness, it was known as gender identity disorder and had the following diagnostic criteria:

A. A strong and persistent cross-gender identification (not merely a desire for any perceived cultural advantages of being the other sex). In children, the disturbance is manifested by four (or more) of the following:

1. Repeatedly stated desire to be, or insistence that he or she is, the other sex.

2. In boys, preference for cross-dressing or simulating female attire; in girls, insistence on wearing only stereotypical masculine clothing.

3. Strong and persistent preferences for cross-sex roles in make-believe play or persistent fantasies of being the other sex.

4. Intense desire to participate in the stereotypical games and pastimes of the other sex

5. Strong preference for playmates of the other sex.

B. Persistent discomfort with his or her sex or sense of inappropriateness in the gender role of that sex.

C. The disturbance is not concurrent with a physical intersex condition.

D. The disturbance causes clinically significant distress or impairment in social, occupational, or other important areas of functioning.
("DSM-IV-TR Diagnostic Criteria for Gender Identity Disorder," 2003)

The next three sections will explain gender dysphoria, its causes and why so many in society are reporting to have it.

Section Two: The Five Main Causes of Gender Dysphoria

This list may be incomplete, but these five causes can explain the source of many cases of gender dysphoria. This section will define and explain these five causes. The causes being: Cluster B personality disorders, internalized shame over sexuality and/or sex non-conforming behavior, trauma, psychosis and autism. It is important to note that there are more reasons for gender dysphoria most likely and these causes can and do overlap.

Cluster B personality disorder refers to these four disorders: antisocial personality disorder, borderline personality disorder, histrionic personality disorder, and narcissistic personality disorder. These disorders have their differences, but what they all result in is the individual having an unstable sense of self. This unstable identity forms a foundation for gender dysphoria to manifest if a person with a cluster B personality disorder experiences life events that trigger discomfort with their natal sex and/or fall victim to the social contagion element of gender dysphoria.

It is unfortunately the case that many people who are homosexual and/or exhibit behavior that is considered atypical of their natal sex are shamed for it. This shame can cause distress and the distress will manifest into gender dysphoria. An effeminate gay man who cannot accept himself for who he is will conceptualize himself as a straight woman and desire to be perceived as such as in the hopes that the shame from society that has been internalized will cease. A similar phenomenon can be

71

observed in lesbians with mannerisms that are typically masculine. This group of people are most likely to become transsexuals, and if they transition as adults in a good frame of mind, they will reap the most relief from gender dysphoria by undergoing social and medical transition. However, many in this group will undergo transition as children and if that intervention were to not occur then they would develop into adulthood as non-conforming homosexuals and not transsexuals in most cases. Gender ideologues have caused great harm to this group as children by making them prime targets to contract socially induced gender dysphoria.

Severe trauma unfortunately can impact one's mental health in many negative ways. Sexual trauma especially so, as it is an extreme violation of one's personhood. There are many psychiatric complications that can form because of this trauma, and gender dysphoria is one. Girls that are molested when they are pre-pubescent or in the early stages of puberty will often feel great distress over the fact that the development of their secondary sexual characteristics will garner unwanted attention from men. These girls will either find or be presented with social and medical transition to prevent such attention. A successful transition can imbue these girls with a male appearance and relieve the distress, but the underlying trauma is not treated by this and merely placated. Men who experience sexual trauma in the homosexual community will be emasculated in these experiences and desire that sexual dynamic and develop dysphoria as a result. While a successful transition can let these men have a sexual dynamic like that of a straight woman, this does not heal the trauma that has been inflicted on them.

Psychosis is a mental disorder where one cannot distinguish what is part of reality and what is not. Some people with psychosis will develop the delusion that they genuinely are the opposite sex trapped in the body of their natal sex. This delusion can be managed with transition, but treatment of the underlying psychosis is what would be most effective in resolving the gender dysphoria.

Autism is an umbrella term that refers to a range of developmental disorders. This, along with the individualized nature of mental health, makes it hard to spot general trends in those with autism, but it is possible to an extent. Many people with autism can easily dissociate and engage in hyper-fixation. Men with autism can dissociate from their physical selves and fixate on the actresses in pornography and other media. As they identify with these women, they will form gender dysphoria as they reattach to an identity that is radically different from their natal sex. Autistic women will dissociate and then hyper-fixate on a new identity as well, but it is usually not because of porn consumption. Autism does not guarantee that one will develop gender dysphoria, and it is usually accompanied by the other causes. Autism also makes one more vulnerable to developing gender dysphoria because of social contagion. It has been observed that there are elevated rates of autism in the group of people who take on a trans identity. Traits of autism are more prevalent in this group as well, even when comparing non-autistic transgender individuals with the general population (Warrier et al, 2020).

The five causes explained in this section are most likely not the only causes of gender dysphoria and are not mutually exclusive. Gender dysphoria is a mental illness and, like all mental illnesses, the root causes can be hard to pinpoint and will vary greatly among individuals. Gender dysphoria can develop because of one's social environment. This is the social contagion element of gender dysphoria known as rapid-onset gender dysphoria, which will be discussed in the next section.

Section Three: The Social Contagion Element: Rapid-Onset Gender Dysphoria

People are products of their environment. While this can lead to improved lives if people find themselves in the right environment, being in the wrong environment can have disastrous effects. Social contagions are one of these disastrous effects, and they are not a new phenomenon. Even before the internet, when eating disorders were discussed in print media, there were droves of young girls and women that reported having the same eating disorders, and now the internet age has only made social contagions more prevalent. Many mental illnesses have increased in diagnosis when prominent social media users have reported to have a mental illness and used social media to document their experience in dealing with it. Gender dysphoria has been one of the illnesses that has spread because of being publicized online. This phenomenon is known as rapid-onset gender dysphoria, or ROGD. ROGD has affected mostly young girls, as they typically experience a greater need for social acceptance than boys and thus are more susceptible to social contagions. ROGD is more likely to form if the person is in a physical environment that pushes gender ideology, like a public school, but social media is so effective at spreading social contagions that one can get ROGD even if their physical environment is hostile to gender ideology. ROGD is easier to develop in people who deal with other factors that can cause gender dysphoria. While ROGD does and can develop quickly, it can also be resolved quickly if the person removes themself from the environment that gave them ROGD.

While ROGD can be resolved quickly, the concerning thing is that many gender ideologues function in a cult-like manner. Once one presents with ROGD, those who inflicted the social contagion onto them will do as much as they can to drive them as far down the path of transition as possible. This is done by having the person with ROGD undergo social transition in secret from friends and family that could voice objections and to then have them undergo medical transition if they have physical access to them or if they can ship the drugs needed to start HRT using the internet to facilitate the transaction. If one with ROGD can go to a physical gender clinic, then they will usually be coached into claiming that they have the traditional diagnosis of gender dysphoria instead of presenting with sudden feelings of discomfort during adolescence or adulthood after they entered a social environment that spreads ROGD.

The rise of ROGD is one of the driving factors that

has helped push gender ideology into its current position. The justification for any collateral or intentional damage of gender ideology is that it will relieve gender dysphoria. The relief of gender dysphoria will supposedly save lives, and the more who suffer with gender dysphoria then the more justified gender ideology will seem to be. This is why gender ideologues will vehemently oppose the idea that gender dysphoria can be a social contagion. They desire to frame gender dysphoria as only the result of society not accepting an individual's trans identity and not the result of a mental illness that develops from internal factors or from their environment as a contagion.

Section Four: Conclusion

Gender dysphoria is a mental illness. There are a variety of causes for this mental illness, but they all stem from complications that are internal or external that prevent one from having a solid identity. There is also the fact that gender dysphoria is a social contagion, and many have presented with a version of the illness called ROGD. The treatment that gender ideologues present is social, medical transition, and societal acceptance of all the tenets of gender ideology. This treatment has created many problems for the individual and society as discussed in previous chapters. Gender ideology should be treated and viewed as the mental illness it is. In order to understand gender dysphoria as a mental illness, then one must understand how the psyche conceptualizes the conflict between what one's natal sex is and what one's perception is. This will be discussed in the next chapter.

Chapter Two: Autogynephilia and Autoandrophilia

Section One: Introduction

This text has established that gender dysphoria is a mental illness where one's perception of their sex conflicts with reality. The psyche must conceptualize this conflict and the result of this are the phenomena of autogynephilia and autoandrophilia. Autogynephilia has been established in the academic community as a fetish where men gain sexual satisfaction from envisioning themselves as women. While this is not entirely untrue, this phenomenon is more complex and applies to women who experience gender dysphoria as well.

Autogynephilia is the desire to perceive oneself as female and/or to be perceived as female by society. This not only applies to physical appearance but the social and sexual functions that are stereotypically unique to women. This can be a fetishist desire, and the inverse applies to women as well. It is also important to note that one can experience autogynephilia and autoandrophilia without experiencing gender dysphoria. This chapter will explain autogynephilia and autoandrophilia in as much detail as possible with the author's current understanding of the phenomena. In general, the understanding of the phenomena is limited, as the academic community has abandoned most, if not all, research into these phenomena as it clearly frames gender dysphoria as a mental illness.

Section Two: Autogynephilia

It was mentioned in the introduction to this chapter that the academic community established the term autogynephilia but only from the perspective of it being a fetishistic desire. This definition of autogynephilia fails to explain how every male who presents with gender dysphoria conceptualizes their identity as a woman. There are homosexual, heterosexual, and bisexual autogynephiles, and this section will explain all instances of autogynephilia along with men that display autogynephilic tendencies.

Homosexual autogynephiles are also known as homosexual transsexuals. They are effeminate gay men that are more comfortable with conceptualizing themselves as straight women rather than gay men. These men also desire the relationship dynamics of heterosexual couples rather than the ones that form in the community of gay men. The way these men portray womanhood is slightly more restrained and realistic than heterosexual autogynephiles but is often far more flamboyant than the average woman would be comfortable with. The gender dysphoria that these men experience usually conforms to the typical diagnosis and starts in early childhood. As such, this group of men is extremely vulnerable to falling to ROGD but can also gain the most resolution of gender dysphoria by undergoing social and medical transition.

Heterosexual autogynephiles display the behaviors that academics studied and named in the mid-twentieth century. This group of men are fetishists that gain sexual gratification from cross-dressing, being acknowledged as women and having sexual experiences as women. Whether they engage in these fantasies alone, with other women, or with men, where they can envision themselves as the women they would want to have sex with, depends on the proclivities of the individual. The sexual fantasies can involve these men being able to have lesbian sex as natal women or having heterosexual interactions with some or all the secondary sexual characteristics of women. Some of these men will even act out fantasies in public settings. Masturbation in women's locker rooms or bathrooms is an example of the most fetishistic behavior that autogynephilic men engage in. Some of these men can compartmentalize their fetish from their identity, while others cannot and experience gender dysphoria because they cannot gain satisfaction from their fetish through living as their natal sex or social and medical transition. It is important to note that much of this fetishism is fueled by pornography, but one can become an autogynephile without the consumption of pornography.

Bisexual autogynephiles are men that are fetishists like heterosexual autogynephiles but are willing to have sexual interactions with men to placate their fetish of being a woman in sexual situations. These men typically will not desire to be straight women in the same way as homosexual autogynephiles, and they are slightly less inclined to engage in the depraved fetishism that heterosexual autogynephiles do, but it is still possible for them to do so.

There are men who have autogynephilic tendencies and live as their natal sex and do not experience gender dysphoria. They will cross-dress and act in an effeminate manner for personal and or sexual gratification in a much more limited fashion than true heterosexual autogynephiles. This group of men are vulnerable to ROGD, as their behavior can be construed to be dysphoria when it is not. These men will engage with stereotypical feminine behavior on limited terms and have a separate and compartmentalized identity as a man for most of their social interactions.

In conclusion, autogynephilia is much more complex than how it was originally described by the academic community. Autogynephiles can be homosexual men that desire to be straight women, heterosexual men that form the identity of a woman to satisfy a sexual fetish, or bisexual men that adopt an identity that takes aspects from both homosexual and heterosexual autogynephiles. There are also men that have autogynephilic tendencies but do not have gender dysphoria; their behavior is too restrained to fall into the realm of being considered an autogynephile. This explanation of autogynephilia encompasses the fetishists that allowed academics to first identify the phenomenon as well as to explain the other more restrained versions of womanhood that some men desire.

Section Three: Autoandrophilia

Autoandrophilia, unlike autogynephilia, was far less an established concept before gender ideologues purged the study of this phenomenon altogether. It is much like autogynephilia, and autoandrophiles can be either homosexual, heterosexual or bisexual. There are also women who display autoandrophilic tendencies. All of these will be explained in this section.

Heterosexuals are the largest group of autoandrophiles. These are straight women that desire to be gay men and to participate in the homosexual community in the same way that gay men do. Whether these women want to engage in homosexual interactions in a completely dominant manner, a completely submissive manner, or a combination of both is dependent on the proclivities of the individual. While many of these women desire to have male genitalia, there are some women who wish to be perceived as homosexual men while keeping their natal genitalia and even their secondary sexual characteristics in some cases. There is a lot of fetishism in the homosexual community, and many of these women will engage with that fetishism to satisfy their autoandrophilia.

Homosexual autoandrophiles are women that would be 'butch' lesbians, but since they experience gender dysphoria, they use medical and social transition to attempt to live their lives as straight men to relieve the dysphoria they feel. They also desire the relationship dynamics typical of heterosexual couples instead of lesbian couples and will desire partners that are and act like straight women instead of lesbians. These autoandrophiles will desire to have a completely male appearance and will want to have sexual interactions in the same manner that natal males can with women. While some of this group seek non-fetishistic relationships, others engage in relationships to satisfy their sexual proclivities alone.

Bisexual autoandrophiles can display some or all

aspects of both homosexual and heterosexual autoandrophiles. They can be more likely to be fetishists, as they can have the highest number and most varied sexual interactions of all autoandrophiles.

Some women will engage with masculine behavior or desire to be perceived as masculine without being true autoandrophiles. These women will typically have childhoods where they rebelled against femininity and acted like stereotypical tomboys. In adulthood, these women accept their natal sex and usually act in a more typical feminine manner but retain some of their tomboyish behavior from childhood.

Autoandrophilia is a phenomenon that needs much more research to be fully understood. What is known is that there are homosexual autoandrophiles that would otherwise be butch lesbians and try to satisfy their desire to be straight men; heterosexual autoandrophiles that desire to be members of the homosexual male community; and bisexual autoandrophiles that can desire both these lifestyles in tandem or gravitate heavily toward one. There are also women that have autoandrophilic tendencies, but it is not a manifestation of gender dysphoria or behavior that should be regarded as autoandrophilia.

Section Four: Conclusion

Autoandrophilia and autogynephilia are phenomena that are vastly understudied, but there is information that has been established. They are similar paraphilias that can be engaged with in a manner that is not necessarily fetishistic but often is. When one experiences gender dysphoria, the psyche will attach to scenarios that relieve it and the result is one feeling the desire to become the opposite sex and to be perceived as such. There are those who engage with the behaviors that are the result of autogynephilia and autoandrophilia but do not have either paraphilia. These concepts have had little study in the academic and medical community and the ideas are discouraged as they challenge the core tenets of gender ideology. Along with the fact that acceptance that these phenomena exist also classifies many of the most ardent gender ideologues as fetishists instead of innocent people finally able to live as their identity. However, acceptance of these phenomena and some understanding of it allows one to have a clear understanding that gender dysphoria is a mental illness. The causes affect the mind, and they result in certain behaviors and thought patterns. This text has established that medical and social transition have negatives that are best to be avoided. However, those who suffer from gender dysphoria will naturally and rightfully want relief from their mental illness. The next chapter will introduce methods to manage gender dysphoria without transition.

Chapter Three: How to Manage Gender Dysphoria without Transition

Section One: Introduction

The medical community has a vested financial interest in portraying transition, especially medical transition, as the only meaningful cure to gender dysphoria (Ugalmugle & Swain, 2023) ("U.S. Sex Reassignment Surgery Market Size, Share & Trends Analysis Report," 2023). The author of this text disagrees with this sentiment and will lay out a method to manage gender dysphoria by understanding basic neuroscience, practicing mindfulness, and cognitive behavioral therapy. To be clear, the methods laid out in the next three sections of this text are not a cure for gender dysphoria. For these methods to be considered a cure, they would have to be standardized, refined and tested with multiple randomized control trials with three groups. A control group that receives only standard psychiatric care; a group that is treated in a manner like what this text will present; and a group that undergoes transition. The author does not recommend that those who have gone through medical transition and are comfortable with the changes to practice this methodology. These methods are for those who experience dysphoria and want relief without going through transition or detransitioners that still have gender dysphoria but have exhausted transition as a method to treat it. It should also be noted that attempting these methods with a biased mindset against the text itself and/or the author will guarantee its ineffectiveness. It also must be emphasized that if you are experiencing severe mental distress and suicidal ideation, these methods are not a replacement for psychiatric care. This methodology is not made by or approved by any licensed medical professional.

Section Two: Using Neuroscience to Manage Gender Dysphoria

The advent of neuroscience has helped make psychology a legitimate field of study. The ability to use scanning technology like MRI and CT has allowed us to gain insight into how the brain functions and how that function affects cognition and behavior. While the brain is an incredibly complex organ, this section will only focus on one part of the brain: The reticular activation system, or RAS.

The RAS is a net-like formation of nerve cells in the brain stem between the brain and the spinal cord. What the RAS does is to filter the thousands of stimuli that you are exposed to and only allow your brain to process the most important ones. Knowledge of this function of the brain can allow you to make it work to your advantage. If I told you to look around your environment for anything red, then you will likely see red objects you did not see before, or if I asked you what the tips of your feet feel like you will become aware of how your feet feel when you most likely were not paying attention to that. If you have gender dysphoria, then you can reduce the severity of your dysphoria by eliminating things from your environment that generate dysphoric thoughts or have the possibility to do so. If you can cease social media use, that would be best in giving your

RAS less dysphoria-inducing thoughts, but filtering topics related to gender and gender ideology from your social media feeds can be immensely helpful as well. Any objects that make you feel dysphoric should be disposed of. If there are people in your physical environment that trigger dysphoric thoughts, then do your best to create healthy boundaries with them so they do not engage in the behavior that triggers dysphoric thoughts, or eliminate contact with them in as friendly a manner as possible.

Eliminating external sources of dysphoric thoughts can only go so far. The mind generates hundreds of thoughts daily, and many can be negative, especially if you suffer from a mental illness like gender dysphoria. Forging the skill of mindfulness is how you can help to further filter dysphoric thoughts. What mindfulness is and how to build it as a skill will be discussed in the next section.

Section Three: How to Build Mindfulness and Use It to Manage Gender Dysphoria Dysphoria

Mindfulness is to be aware of not only the environment and its changes but yourself and the fluctuations in thoughts and feelings you experience from day to day. Building the skill of mindfulness is a lifelong process, and meditation is the primary method to become more mindful. This section offers a guide on how to meditate as well as how to be mindful and how to use it to deal with negative thoughts, including those associated with gender dysphoric.

Many think of meditation as this ancient and mystical practice that only a select few can unlock the secrets of. This is false, as anybody can meditate, and it is easy to integrate meditation into your daily life. There are many guided meditations available as video and audio files online and apps on smartphones. Unguided meditation can be practiced as well. The method to do so is to set a timer for at least five minutes and focus on your breath and where you feel it the most. During these sessions, you do not put effort into breathing, you merely make it the sole thing you focus on and when other thoughts arise, acknowledge them and let them pass. Meditation is not about achieving a perfectly clear mind or entering a state of perfect serenity, but it is a practice where you acknowledge that you do not control your stream of consciousness and are merely a steward of it. Setting aside five minutes to meditate daily will give you awareness of your thoughts. Once you are aware of your thoughts then you can let them pass as easily as sounds and sights can from your awareness. Being able to do this is what it means to build the skill of mindfulness, and it is a lifelong practice. However, one can make significant gains in the skill of mindfulness with just a few weeks of consistent practice. Velvet Room Publishing has released a beginner's guide to meditation alongside this text.

Being mindful can only get you so far in managing your dysphoric thoughts. Negative thoughts can often be overwhelming and difficult to rationalize. The next section will introduce a method to define negative thoughts, rationalize, and then counter them. This will work for all negative thoughts and not just those brought about because of gender dysphoria.

Section Four: Using Cognitive Behavioral Therapy to Manage Gender Dysphoria

Cognitive behavioral therapy, or CBT, is a method of psychiatric care that does not involve the use of drugs; instead, it relies on changing the patient's thought patterns. CBT allows you to identify your negative thoughts, discard them and then put positive ones in their place. David D. Burns, MD, lays out this method in his book The Feeling Good Handbook and this text will cite his methods and show how they can be applied to negative thoughts relating to gender dysphoria.

The first step in being able to counter negative thoughts is to identify them. The Feeling Good Handbook explains that negative thoughts can fall into ten categories:

1.   All-or-nothing thinking
2.   Overgeneralization
3.   Mental filter
4.   Discounting the positive
5.   Jumping to conclusions (Mind reading and Fortune-telling)
6.   Magnification
7.   Emotional reasoning
8.   Should statements
9.   Labeling
10.   Personalization and blame

To better identify how your negative thoughts fit into these categories, a summarization of each is provided below:

1 .All-or-nothing thinking is when you only view everything as purely black and white. A scenario can only end in either complete success or total failure and often little mistakes will become total failures in your mind due to this negative thought pattern.

2. To overgeneralize is to take a single negative event and assume that it means you can only experience negative events.

3. Mental filtering is when you actively seek out negative thoughts and discard positive ones.

4. Discounting the positive is when you take positive thoughts and refute them into a negative thought.

5. Jumping to conclusions comes in two forms: Mind reading—when you assume someone else thinks negatively of you, and Fortune-telling—where you predict your life will turn out poorly.

6. Magnification is when you exaggerate the severity of your problems and understate the positives in your life.

7. Emotional reasoning is when you use your feelings to justify self-destructive thoughts and behaviors instead of having a well-thought-out, logical explanation.

8. A should statement is when you imply shame to yourself over a shortcoming. This is done by starting the thought with the words I should, or I shouldn't.

9. Labeling is a more extreme form of all-or-nothing thinking where you label yourself in a way that can only lead to negative thoughts.

10. Personalization and blame are when you take responsibility for a situation out of your controcontrol and then ascribe blame as it is or seen as a negative scenario. (Burns, 1999)

Now that you can define your negative thoughts, you can filter them and change them into positives. This is best done by handwriting on a page with three columns. The first column lists negative thoughts you have, the second column lists definitions of what the negative thoughts are, and the third column lists the refutations that turn these negative thoughts into positive ones. The Feeling Good Handbook goes into detail on how to do this along with other methods of CBT. Reading The Feeling Good Handbook is recommended, but this can be performed with the information in this text alone. Velvet Room Publishing will also release a video outlining how to do this exercise using thoughts related to gender dysphoria.

Using CBT can help manage all negative thoughts, not just those related to gender dysphoria. The methods of CBT described by David D. Burns, MD in The Feeling Good Handbook have been proven to be effective against a variety of mental illnesses. While there is not much documentation on using CBT to treat gender dysphoria in particular, the same methods can and most likely do apply, especially if used in conjunction with other forms of psychiatric care.

Section Five: Conclusion

Gender dysphoria is a mental illness, and as such it should be treated like any other mental illness. The methodology presented here involves understanding neuroscience and that isolating external sources of dysphoric thoughts can lessen dysphoria; building the skill of mindfulness to allow dysphoric thoughts to feel less intense and pass easier; and using cognitive behavioral therapy or CBT to define negative and dysphoric thoughts and transform them into positive ones.

Following this methodology requires no use of drugs or the performance of any surgeries, unlike medical transition. This is not to say that this methodology is a legitimate medical treatment or a replacement for professional psychiatric care, especially if you are suicidal or having a severe mental health episode. It is a low-cost, no-risk method to relieve dysphoric thoughts by strengthening the mind's ability to deal with them.

Benjamin Bode

Afterword

Thank you for reading this text. I hope if you read the entire text closely or merely skimmed through it that it was useful and informative to you. Gender ideology's concerning encroachment on society compelled the author to write this text. While the author asserts that gender ideologues have harmed society, he does not harbor ill-will toward them. This text was authored to explain gender ideology in the context of reality and to provide methodology of how to deal with its negative effects.

This text was intended to explain the core tenets of gender ideology and refute them with clear, concise language that the average person can understand. This text went into detail on the process of transition—some of the more gruesome complications can only be explained by diving into firsthand accounts of those who received those surgeries. This is most true for vaginoplasty and phalloplasty. If you desire to have the deepest understanding of gender ideology that you can, then you must do your own research, not only into the established medical literature but firsthand accounts from gender ideologues and those who fell victim to gender ideology and were then left to speak out against it.

References

Al-Enezi, A., Kehinde, E. O., Behbehani, A., & Sheikh, Z. A. (2007). Luteinizing Hormone-Releasing Hormone Analogue-Induced Cataract in a Patient with Prostate Cancer. Medical Principles and Practice, 16(2), 161–163. https://doi.org/10.1159/000098373

Baillargeon, J., Snih, S. a. S. A., Raji, M., Urban, R. J., Sharma, G., Sheffield-Moore, M., López, D. S., Baillargeon, G., & Kuo, Y. F. (2016). Hypogonadism and the risk of rheumatic autoimmune disease. Clinical Rheumatology, 35(12), 2983–2987. https://doi.org/10.1007/s10067-016-3330-x

Bilateral or double mastectomy: What to expect and recovery. (n.d.). City of Hope. https://www.cancercenter.com/cancer-types/breast-cancer/treatments/surgery/double-mastectomy

Biggs, M. (2021). Revisiting the effect of GnRH analogue treatment on bone mineral density in young adolescents with gender dysphoria. Journal of Pediatric Endocrinology & Metabolism/Journal of Pediatric Endocrinology and Metabolism, 34(7), 937–939. https://doi.org/10.1515/jpem-2021-0180

Biggs, M. (2022). Suicide by Clinic-Referred Transgender Adolescents in the United Kingdom. Archives of Sexual Behavior, 51(2), 685–690. https://doi.org/10.1007/s10508-022-02287-7

Breast reconstruction after breast cancer: It's not one-size-fits-all. (n.d.). City of Hope. https://www.cancercenter.com/community/blog/2021/07/breast-reconstruction-after-breast-cancer

Burke, S., Manzouri, A., & Savić, I. (2017). Structural connections in the brain in relation to gender identity and sexual orientation. Scientific Reports, 7(1). https://doi.org/10.1038/s41598-017-17352-8

Burns, David D.(1999). The Feeling Good Handbook. New York; Toronto, Ont.: Plume.

Chen, D., Berona, J., Chan, Y., Ehrensaft, D., Garofalo, R., Hidalgo, M. A., Rosenthal, S. M., Tishelman, A. C., & Olson-Kennedy, J. (2023). Psychosocial Functioning in Transgender Youth after 2 Years of Hormones. New England Journal of Medicine, 388(3), 240–250. https://doi.org/10.1056/nejmoa2206297

Crane Center for Transgender Surgery. (2023, August 11). Non-Binary Surgery - Crane Center for Transgender Surgery. https://cranects.com/non-binary-surgery/

Dallas, K., Kuhlman, P., Eilber, K. S., Scott, V., Anger, J. T., & Reyblat, P. (2021). Mp04-20 rates of psychiatric emergencies before and after gender affirming surgery. The Journal of Urology 206(Supplement 3). https://doi.org/10.1097/ju.0000000000001971.20

Dhejne, C., Lichtenstein, P., Boman, M., Johansson, A. L., Långström, N., & Landén, M. (2011). Long-term follow-up of transsexual persons undergoing sex reassignment surgery: Cohort study in Sweden. PloS One, 6(2), e16885. https://doi.org/10.1371/journal.pone.0016885

Donovan Cleckley on X: "'Every single child, or adolescent, who was truly blocked at Tanner stage 2 [by age 11] has never experienced orgasm. I mean, it's really about zero.' - Marci Bowers, 'Trans & Gender Diverse Policies, Care, Practices, and Wellbeing' (2022) https://t.co/u7ohsal5JT" / X. (n.d.). X (Formerly Twitter). https://twitter.com/DonovanCleckley/status/15216255 18394773505?s=20

DSM-IV-TR. Diagnostic Criteria for Gender Identity Disorder. (2003). Psychiatric News, 38(14), 32. https://doi.org/10.1176/pn.38.14.0032

Ek, M., Roth, B., Ekström, P., Valentin, L., Bengtsson, M., & Ohlsson, B. (2015). Gastrointestinal symptoms among endometriosis patients—A case-cohort study. BMC Women's Health, 15(1). https://doi.org/10.1186/s12905-015-0213-2

Fariba, N., Soheilirad, Z., & Haghshenas, Z. (2019). The influence of gonadotropin-releasing hormone agonist treatment on thyroid function tests in children with central idiopathic precocious uberty. Medicinski Arhiv, 73(2), 101. https://doi.org/10.5455/medarh.2019.73.101-103

Feminizing hormone therapy - Mayo Clinic. (2023, December 2). https://www.mayoclinic.org/tests-procedures/feminizing-hormone-therapy/about/pac-20385096

Finlayson, C., Johnson, E. K., Chen, D., Dabrowski, E., Gosiengfiao, Y., Campo-Engelstein, L., Rosoklija, I., Jacobson, J. D., Shnorhavorian, M., Pavone, M. E., Moravek, M. B., Bonifacio, H. J., Simons, L., Hudson, J., Fechner, P. Y., Gomez-Lobo, V., Kadakia, R., Shurba, A., Rowell, E. E., & Woodruff, T. K. (2016). Proceedings of the working group session on fertility preservation for individuals with gender and sex diversity. Transgender Health, 1(1), 99–107. https://doi.org/10.1089/trgh.2016.0008

Gluck, G. (2022, June 17). Top trans medical association collaborated with castration, child abuse fetishists - Reduxx. Reduxx. https://reduxx.info/top-trans-medical-association-collaborated-with-castration-child-abuse-fetishists/

Guillamón, A., Junqué, C., & Gómez-Gil, E. (2016). A review of the status of brain structure research in transsexualism. Archives of Sexual Behavior, 45(7), 1615–1648. https://doi.org/10.1007/s10508-016-0768-5

Hammar, O., Roth, B., Bengtsson, M., Mandl, T., & Ohlsson, B. (2013). Autoantibodies and gastrointestinal symptoms in infertile women in relation to in vitro fertilization. BMC Pregnancy and Childbirth, 13(1). https://doi.org/10.1186/1471-2393-13-201

Hayes, P. (2017). Commentary: Cognitive, emotional, and psychosocial functioning of girls treated with pharmacological puberty blockage for idiopathic central precocious puberty. Frontiers in Psychology, 8. https://doi.org/10.3389/fpsyg.2017.00044

Heino, E., Fröjd, S., Marttunen, M., & Kaltiala, R. (2023). Transgender identity is associated with severe suicidal ideation among Finnish adolescents. International Journal of Adolescent Medicine and Health, 35(1), 101–108. https://doi.org/10.1515/ijamh-2021-0018

Hoffman, M., MD. (2023, August 7). The different types of hysterectomy and their benefits. WebMD. https://www.webmd.com/women/hysterectomy

Horbach, S. E., Bouman, M., Smit, J. M., Özer, M., Buncamper, M. E., & Mullender, M. G. (2015). Outcome of vaginoplasty in male-to-female transgenders: A systematic review of surgical techniques. OUP Academic. https://doi.org/10.1111/jsm.12868

Hough, D., Bellingham, M., Haraldsen, I., McLaughlin, M., Robinson, J. E., Solbakk, A., & Evans, N. P. (2017). A reduction in long-term spatial memory persists after discontinuation of peripubertal GnRH agonist treatment in sheep. Psychoneuroendocrinology, 77, 1–8. https://doi.org/10.1016/j.psyneuen.2016.11.029

Huang, C., Doole, R., Wu, C., Huang, H., & Yi, C. (2019). Culture-related and individual differences in regional brain volumes: A cross-cultural voxel-based morphometry study. Frontiers in Human Neuroscience, 13. https://doi.org/10.3389/fnhum.2019.00313

Is a keyhole mastectomy the best option for me? | McLean Clinic. (2022, April 5). https://www.topsurgery.ca/blog/keyhole-mastectomy-best-option

Klink, D., Caris, M. G., Heijboer, A. C., Van Trotsenburg, M., & Rotteveel, J. (2015). Bone mass in young adulthood following Gonadotropin-Releasing Hormone analog treatment and Cross-Sex hormone treatment in adolescents with gender dysphoria. Journal of Clinical Endocrinology & Metabolism, 100(2), E270–E275. https://doi.org/10.1210/jc.2014-2439

Kurth, F., Gaser, C., Sánchez, F. J., & Lüders, E. (2022). Brain sex in transgender women is shifted toward gender identity. Journal of Clinical Medicine, 11(6), 1582. https://doi.org/10.3390/jcm11061582

Loree, J., Burke, M. S., Rippe, B., Clarke, S., Moore, S. H., & Loree, T. R. (2020). Transfeminine gender confirmation surgery with penile inversion vaginoplasty: An initial experience. Plastic and Reconstructive Surgery. Global Open, 8(5), e2873. https://doi.org/10.1097/gox.0000000000002873

Lüders, E., Sánchez, F. J., Gaser, C., Toga, A. W., Narr, K. L., Hamilton, L. S., & Vilain, É. (2009). Regional gray matter variation in male-to-female transsexualism. NeuroImage, 46(4), 904–907. https://doi.org/10.1016/j.neuroimage.2009.03.048

Masculinizing hormone therapy - Mayo Clinic. (2023, December 2). https://www.mayoclinic.org/tests-procedures/masculinizing-hormone-therapy/about/pac-20385099

Masculinizing surgery - Mayo Clinic. (2022, November 4). https://www.mayoclinic.org/tests-procedures/masculinizing-surgery/about/pac-20385105

Mueller, S. C., De Cuypere, G., & T'Sjoen, G. (2017). Transgender research in the 21st century: A selective critical review from a neurocognitive perspective. The American Journal of Psychiatry, 174(12), 1155–1162. https://doi.org/10.1176/appi.ajp.2017.17060626

Nguyen, H. B., Loughead, J., Lipner, E., Hantsoo, L., Kornfield, S. L., & Epperson, C. N. (2018). What has sex got to do with it? The role of hormones in the transgender brain. Neuropsychopharmacology, 44(1), 22–37. https://doi.org/10.1038/s41386-018-0140-7

Nitkin, K. (2018, April 23). First-ever penis and scrotum transplant makes history at Johns Hopkins. Johns Hopkins Medicine. https://www.hopkinsmedicine.org/news/articles/2018/04/first-ever-penis-and-scrotum-transplant-makes-history-at-johns-hopkins

Nokoff, N., Scarbro, S., Moreau, K. L., Zeitler, P., Nadeau, K. J., Reirden, D., Juarez-Colunga, E., & Kelsey, M. M. (2021). Body composition and markers of cardiometabolic Health in transgender youth on gonadotropin-releasing hormone agonists. Transgender Health, 6(2), 111–119. https://doi.org/10.1089/trgh.2020.0029

Osborn, C. O. (2018, September 18). Phalloplasty: gender confirmation surgery. Healthline. https://www.healthline.com/health/transgender/phallo plasty#risks-and-complications

Overview of feminizing hormone therapy | Gender Affirming Health Program. (n.d.). https://transcare.ucsf.edu/guidelines/feminizing-hormone-therapy

Peritoneal PullThrough Vaginoplasty FAQs | mozaic. (n.d.). Mozaic. https://www.mozaiccare.net/peritoneal-pullthrough-vaginoplasty

Phalloplasty for gender affirmation. (2023, May 3). Johns Hopkins Medicine. https://www.hopkinsmedicine.org/health/treatment-tests-and-therapies/phalloplasty-for-gender-affirmation

Pharmacovigilance Working Party (PhVWP). (2012). Patient health protection monthly report Issue number: 1112. In Pharmacovigilance Working Party (PhVWP) December 2011 (p. 1) [Report]. https://www.ema.europa.eu/en/documents/report/mo nthly-report-pharmacovigilance-working-party-phvwp-december-2011-plenary-meeting_en.pdf

Planned Parenthood. (n.d.). Effects of masculinizing hormone therapy. https://www.plannedparenthood.org/uploads/filer_public/15/38/15388d56-b877-4829-b28a-c88f9d5ea11f/2043_effects_of_masculinizing_hormone_therapy_0417.pdf

Professional, C. C. M. (n.d.). Masculinizing hormone therapy. Cleveland Clinic. https://my.clevelandclinic.org/health/treatments/22322-masculinizing-hormone-therapy

Professional, C. C. M. (n.d.). Metoidioplasty. Cleveland Clinic. https://my.clevelandclinic.org/health/procedures/21668-metoidioplasty

Reddit.com/r/medicine. (2023, August 3). The Chen 2023 Paper Raises Serious Concerns About Pediatric Gender Medicine Outcomes. Reddit.com. https://archive.fo/lDlwk#selection-1481.0-1481.84

Rice, Cari N., Grundy, Virginia. Howard, Cherie H. (2006, September 22).. Complications of Hysterectomy. U.S. Pharmacist. https://www.uspharmacist.com/article/complications-of-hysterectomy

Rocca, W. A., Grossardt, B. R., Shuster, L. T., & Stewart, E. A. (2012). Hysterectomy, oophorectomy, estrogen, and the risk of dementia. Neurodegenerative Diseases, 10(1–4), 175–178. https://doi.org/10.1159/000334764

Savić, I., & Arver, S. (2011). Sex dimorphism of the brain in Male-to-Female transsexuals. Cerebral Cortex, 21(11), 2525–2533. https://doi.org/10.1093/cercor/bhr032

Schemmel, A The National Desk. (n.d.). FDA warns puberty blocker may cause brain swelling, vision loss in children. WNWO. https://nbc24.com/news/nation-world/fda-warns-puberty-blocker-may-cause-brain-swelling-vision-loss-in-children-rachel-levine

Singh, D., Bradley, S. J., & Zucker, K. J. (2021). A follow-up study of boys with gender identity disorder. Frontiers in Psychiatry, 12. https://doi.org/10.3389/fpsyt.2021.632784

Smith, J. A., & Urry, R. L. (1985). Testicular histology after prolonged treatment with a gonadotropin-releasing hormone analogue. The Journal of Urology/the Journal of Urology, 133(4), 612–614. https://doi.org/10.1016/s0022-5347(17)49110-5

The side effects of gonadotropin releasing hormone analog (diphereline) in treatment of idiopathic central precocious puberty. (2013). PubMed. https://pubmed.ncbi.nlm.nih.gov/23456583/

Tordoff, D. M., Wanta, J. W., Collin, A., Stepney, C., Inwards-Breland, D. J., & Ahrens, K. (2022). Mental health outcomes in transgender and nonbinary youths receiving Gender-Affirming Care. JAMA Network Open, 5(2), e220978. https://doi.org/10.1001/jamanetworkopen.2022.0978

Treatment Risks and side effects of feminizing hormone therapy. (n.d.). https://mydoctor.kaiserpermanente.org/mas/structured-content/Treatment_Risks_and_Side_Effects_of_Feminizing_Hormone_Therapy.xml?co=%2Fregions%2Fmas

Ugalmugle, S., & Swain, R. (2023, February 9). Sex Reassignment Surgery Market size to exceed $1.95Bn by 2032. Global Market Insights Inc. https://www.gminsights.com/pressrelease/sex-reassignment-surgery-market

U.S. Sex Reassignment Surgery Market Size, Share & Trends Analysis Report By gender transition (female-to-male, male-to-female), by procedure (mastectomy, vaginoplasty, scrotoplasty, hysterectomy, phalloplasty), and Segment Forecasts, 2023 - 2030. (2023, August 14). https://www.grandviewresearch.com/industry-analysis/us-sex-reassignment-surgery-market

Vaginoplasty procedures, complications and aftercare | Gender Affirming Health Program. (n.d.). https://transcare.ucsf.edu/guidelines/vaginoplasty

Wang, Y., Khorashad, B. S., Feusner, J. D., & Savić, I. (2021). Cortical gyrification in transgender individuals. Cerebral Cortex, 31(7), 3184–3193. https://doi.org/10.1093/cercor/bhaa412

Warrier, V., Greenberg, D. M., Weir, E., Buckingham, C., Smith, P., Lai, M., Allison, C., & Baron-Cohen, S. (2020). Elevated rates of autism, other neurodevelopmental and psychiatric diagnoses, and autistic traits in transgender and gender-diverse individuals. Nature Communications, 11(1). https://doi.org/10.1038/s41467-020-17794-1

Yeshaya, A., Kauschansky, A., Orvieto, R., Varsano, I., Nussinovitch, M., & Ben-Rafael, Z. (1998). Prolonged vaginal bleeding during central precocious puberty therapy with a long-acting gonadotropin-releasing hormone agonist. Acta Obstetricia Et Gynecologica Scandinavica, 77(3), 327–329. https://doi.org/10.1034/j.1600-0412.1998.770314.x

Zubiaurre-Elorza, L., Junqué, C., Gómez-Gil, E., Segovia, S., Carrillo, B., Rametti, G., & Guillamón, A. (2012). Cortical thickness in untreated transsexuals. Cerebral Cortex, 23(12), 2855–2862. https://doi.org/10.1093/cercor/bhs267